Introduction to Primate Behavior

Nancy C. Collinge

University of Alberta

KENDALL/HUNT PUBLISHING COMPANY
4050 Westmark Drive   Dubuque, Iowa 52002

Cover photo: A kin group of Japanese macaques.
           Courtesy of Dr. Linda Fedigan.

Copyright © 1993 by Kendall/Hunt Publishing Company

Library of Congress Catalog Card Number: 93-78221

ISBN 0-8403-8569-2

Printed in the United States of America
10   9   8   7   6   5   4   3   2

Contents

Preface

I sincerely hope that this volume fulfills my primary objective of providing straight-forward, easily-digested information about the world of nonhuman primates that will pique the intellectual interest of university students. Most people enjoy watching and reading about animals, however, having spent time observing monkeys, I feel that nonhuman primates deserve special treatment. Along with the introductory chapters describing what it is to be a primate and how we evolved, and the chapters delineating the life-way patterns of a representative number species from each taxonomic group, I devoted over half of the book to a discussion of their impressive social and intellectual abilities. This brings me to another, more selfish objective in writing this book, which is not only to pass on my deep and abiding interest in the animals, but to voice my concerns for the future of our closest living relatives. Certainly, many modern zoos are providing their primate groups with the best possible environment under captive conditions and are important facilities for breeding endangered species. However, it seems sad that so many primate species soon may not be able to roam freely in their tropical forest habitats. It is the youth of the world, the concerned university students, that will have to be the salvation of primates and other endangered species. Any contribution to conservation measures, however small, will help to save what little of the primate habitat that is left. It is my hope that even a little of my fascination for these unique animals rubs off on the readers.

There are a number of individuals who helped and encouraged me at every stage of the writing. I would especially like to acknowledge the contribution of Dr. Linda Fedigan, who not only determined the course of my anthropological studies by her fascinating lectures on primate behavior, but who guided me through my academic career. Her technical input and editorial advice was critical to the completion of this volume. The contribution of my friends in the Anthropology Department at the University of Alberta, in Edmonton was also invaluable. Ruben Kaufman, Francois LaRose and Katie McKinnon were most generous with their photos, as well as their useful suggestions on how to proceed. I would like to thank the Department of Anthropology at the University of Alberta for the use of their facilities and office space during the time I was writing and not officially attending classes or lecturing. Brian Keating, the Education Officer of the Calgary Zoological Society deserves special mention for his willingness to let me "raid" their catalogue of photographs. Dr. Russell A. Mittermeier, President of Conservation International, generously provided photographs as well as current information concerning conservation priorities and the measures being taken to protect nonhuman primates and their habitats. The marvellous photographs supplied by Karen Dickey reminded me of the crucial role in my career played by the Japanese macaques at the South Texas Primate Observatory. If the illustrations in this volume are slightly biased towards Japanese macaques, it is only because of their immense contibution to my commitment to nonhuman primates. However, in the final analysis, it was my friends, particularly Joan Davison, who made life bearable while struggling to be creative and "erudite." Thanks!

Part I

Introduction to the World of Primates

Chapter 1

General Information

Why Study Nonhuman Primates?

One question often put to those who study primate behavior is "I hear that you spend your time watching monkeys behave—so what?" These introductory paragraphs have been written to provide some of the reasons why scholars from 325 B.C. to the present have spent their valuable time studying the biology and behavior of nonhuman primates.

From historical times to the beginning of the 20th century, the principle rationale for studying other primates was that they are so similar to ourselves in appearance. Their striking similarity to humans was first noted by Greek scholars. This inspired the Greeks to study the anatomy of monkeys and apes many centuries before Carolus Linnaeus, the noted Swedish naturalist, classified monkeys and apes along with humans in the "primary" order of living things. However, this physical resemblance to us has been a mixed blessing, as there has been a tendency for scientists to conduct primate studies with the explicit aim of applying the results to the human case, rather than viewing other primates as uinque species in their own right (Richard, 1985). In the last few decades, this situation has been ameliorated, and the discipline of primatology has expanded its horizons to include investigations of the ecological, behavioral, social and cognitive aspects of an ever-increasing number of primate species.

Primatology is a relatively young discipline, only coming into its own in the 1950s after World War II. Its mandate now includes every facet of the nonhuman primate world, thus covering the classification, paleontology, evolution, biology, ecology, and behavior of our closest living relatives. Because the mandate is so all-encompassing, it is a truly interdisciplinary science, practiced by anthropologists, psychologists, zoologists, and biomedical researchers, who focus on every possible aspect of primate life.

Anthropology

Not surprisingly, anthropologists, with their emphasis on the study of human-kind, first put their research energy into the study of nonhuman primates as reflections of the lifeway patterns of early human society. It was felt that the study of the behavior patterns and social organization of monkeys and apes, particularly chimpanzees whose genetic and biochemical makeup is so close to ours, would allow us a glimpse of hominid life in Africa four million years ago. However, it has since become clear that the variability in behavior patterns and social organization that is found between and within primate species precludes the use of any one species as a suitable model for early hominids. In any case, such models were merely analogies and were meant to be treated as such.

Given the penchant of anthropologists to study the social systems of human cultures, it was to be expected that a number of primatologists trained in the discipline would concentrate on the study of nonhuman primate social patterns. The diversity and complexity found in the social organizations of monkeys and apes has not only increased our awareness of their social sophistication, but has provided a good comparative basis for understanding human social behavior.

Before the 1980s, anthropologists were concerned primarily with the evolution of human behavior, and research often focused on continuities and discontinuities in behavior from nonhuman primates to humans. In recent years, primatologists have expanded their areas of interest to include ecological and demographic studies. Now the field of behavioral ecology has become a major area of research, with the emphasis on identifying ecological principles that would explain the behavioral variability between different species as well as the individual variation among members of the same species (DeVore, 1990).

Psychology

The primary focus of psychologists is the study of mental processes such as learning, memory, and overall intelligence in nonhuman primates, in an effort to improve the knowledge of our own cognitive processes. Since it is often impossible or unethical to carry out controlled laboratory experiments on human subjects, animals on our evolutionary line have become acceptable substitutes for psychological testing. However, there is an enormous gap between the level of cognitive functioning in animals and humans, and great care must be taken in extrapolating the use of similar cognitive mechanisms to produce seemingly similar behavioral results.

In the early 1930s C.R. Carpenter, a comparative psychologist, was the first man to successfully study primates in the field. He developed a systematic way of studying the behavior of monkeys that is still in use today. His methods for the accurate description and classification of animal behavior patterns have been used successfully by clinical psychologists in behavioral studies of very young children and mentally handicapped adults and children who are unable to communicate verbally (Blurton-Jones, 1972).

One very important area of primate research that has belonged solely to the psychologists is the field of ape language studies. This body of research has shown that chimpanzees, gorillas and orangutans can use both gestural (in the form of modified American Sign Language) and artificial languages that enable them to communicate with their human trainers. Although this research reached its peak in the 1970s, dedicated psychologists, such as Roger Fouts and Francine Patterson, continue to be involved in the language training of their ape subjects.

Psychologists have demonstrated the impressive abilities of monkeys and apes to master all types of problem-solving tasks in the laboratory. In the 1980s, the growing awareness of the complexity of the social life of nonhuman primates, and the extent and sophistication of their social behavior in the wild, has increased the attention being directed towards the study of social intelligence. The field of social cognition in nonhuman primates promises to be one of the most exciting fields open to researchers in the future.

Zoology

In contrast to the tradition of laboratory research conducted by psychologists, zoologists have contributed greatly to our knowledge of primates in their natural habitat. It is one of the priorities of the biologist to clarify principles pertaining to the evolution of social behavior and social organization in animal species (Cheney et al., 1987). Zoologists have investigated the ecological and demographic factors that ultimately result in a given type of social organization, and have even questioned why primates live in social groups at all.

The introduction of **sociobiological** theory in the 1970s led to a re-examination of earlier studies of animal behavior in the light of modern evolutionary theory, and there has been a dramatic increase in research specifically directed towards determining the adaptive significance for social behavior. As a result, a number of sociobiological principles have been articulated that have greatly influenced the whole field of animal behavior research in general, and nonhuman primates in particular. Kin selection, reciprocal altruism, and parental investment theories are the primary examples of these principles that have been tested on primates. One of the main benefits of an evolutionary perspective has been that primatologists were forced to look at species in their natural habitat and to ask important questions about the function of social behavior. Still, it is clear that it is important not to lose sight of the proximate or immediate factors that affect behavior.

Biomedical Research

Although this book is directed primarily towards a discussion of primate behavior, it would be an oversight not to mention the important role that nonhuman primates have played in biomedical research. Their anatomical, molecular, and biochemical similarity to us has forced monkeys and apes into the role of preferred subjects for testing the efficacy of new drugs, for perfecting surgical procedures, and for the discovery of vaccines to combat communicable diseases. The enormous contribution of these animals to our health has forced us to tread a fine line between our concerns for the conservation of primate species, such as rhesus macaques and chimpanzees, and concerns for the welfare of human populations.

In summary, it is clear that there are a number of very important reasons for studying primate behavior, many of which are selfish, in that the careful use of the results of these studies can further the understanding of our own behavior and biology. However, another extremely urgent reason for continuing to study primates is the fact that over 50% of primate species are in some sort of jeopardy, either threatened, endangered or on the verge of extinction primarily due to loss of habitat. More than 90% of primate species live in the tropical forests of Africa, Asia, and Central and South America, and as these forests disappear, so do the primates. Unless the destruction of these forests is severely constrained, which at this point seems unlikely, many primate species will no longer exist in the wild by the year 2010 (Mittermeier and Cheney, 1987). Thus, it is imperative that we learn as much as possible about the lifeways of as many species as possible. This will add to our fund of knowledge of primates, but will also improve our ability to successfully raise endangered species in captivity, in case they can no longer be sustained in the wild.

The science of primatology is alive and well in the last decade of the 20th century. Researchers from many countries continue to provide valuable data from both field and laboratory studies. It is hoped that this situation will persist. Aside from the satisfaction of adding to the body of scientific information on our closest living relatives, one of the very important rewards of this type of research is the enjoyment derived from watching these wonderful creatures. Each species provides us with a never-ending source of wonder, with its own unique adaptations and characteristics, and yet still possessing so many qualities that remind us of ourselves.

The Taxonomy of the Primates

Although literally speaking, **taxonomy** is "the science of the classification of organisms including the principles, procedures, and the rules of classification" (Nelson and Jurmain, 1991:610), the term is often used to refer to the classification of a group of living organisms. The first comprehensive classification of living plants and animals was compiled by Carolus Linnaeus, a noted Swedish naturalist. In his book, *Systema Naturae*, published in 1758, he categorized the animal kingdom into large groupings which he subdivided into smaller and smaller units based on finer and finer distinctions. He devised the binomial system of scientific nomenclature in which every genus was given a Latin name, followed by a species name referring either to a description of the organism or possibly to the name of the scientist who discovered the species. This system has withstood the test of time and is still in use today.

The Order Primates, which includes lemurs, monkeys, apes and humans, was originally constructed on the basis of similarities in appearances and anatomy rather than biological relationships. Although the Linnaeian system forms the foundation for the present biological system, and the Order Primates still includes these forms, classifications are now based on **phylogenetic** relationships. It is interesting to note that, despite the difference in the basic philosophy underlying the classification, there is an amazing similarity in the categorization designed by Linnaeus and the modern one shown in Table 1.1 of this volume.

Since classification systems are designed by the scientists who use them, it is not surprising that no one primate taxonomy is universally accepted. The basic framework of the primate taxonomy shown in Table 1.1 is a combination of the traditional Linnaeian system, and the recent addition of the Suborder Tarsiodea, to the Suborders Prosimii and Anthropoidea. The tarsiers were formerly grouped with the Suborder Prosimii, primarily because of their small size, **nocturnal** life-style and superficial physical similarity to prosimians. The presence of ancestral tarsiers, along with the early prosimians, in North America and Europe approximately 53 million years ago was also a factor in classifying the two primate forms in the same suborder. However, the tarsier, with the anatomical

6

Table 1.1 The Taxonomy of Living Primates

Order
Suborder
Infraorder
Superfamily
Family—Commmon Names of Representative Species
Subfamily—Common Names of Representative Species
Primates
Prosimii
Lemuriformes
Lemuroidea
Lemuridae—Ring-tailed lemurs, brown lemurs, hapalemurs
Cheirogaleidae—Mouse lemurs, dwarf lemurs etc.
Indriidae—Indri, sifaka
Lepilemuridae—Gentle lemurs, sportive lemurs, etc.
Daubentoniodea
Daubentonidae—Aye-aye
Lorisiformes
Lorisoidea
Lorisidae
Lorisinae—Slender loris, slow loris. potto
Galaginae—Galago, bushbaby
Tarsioidea
Tarsiidae—Tarsiers
Anthropoidea
Platyrrhini
Ceboidea
Callitrichidae—Marmosets and tamarins
Cebidae—Cebus, squirrel, spider, titi, and howler monkeys
Catarrhini
Cercopthecoidea
Cercopithecidae
Cercopithecinae—Macaques, baboons, guenons, etc.
Colobinae—Colobus monkeys, langurs, etc.
Hominoidea
Hylobatidae—Gibbons
Pongidae—Gorillas, orangutans, chimpanzees
Hominidae—Humans

Source: Adapted from Primate Societies 1987

characteristics of both prosimians and anthropoids, has presented a problem for taxonomists.

The tarsiers resemble the prosimians in their nocturnal life-style, the unfused lower jaw, the toilet claws on the second and third digits of the foot and the possession of multiple nipples. On the other hand, they are very anthropoid-like in the completeness of the bony eye-socket, features of the inner-ear anatomy and the system of blood supply to the brain. One of the primary physical traits that separates the tarsiers from the other prosimians is the possession of a dry nose, as opposed to the wet nose common to lorises and lemurs. This has prompted many taxonomists to use this nasal characteristic as a means of dividing the Order Primates into the Suborders Strepsirhini, the wet nosed lemurs and lorises, and Haplorhini, the dry nosed tarsiers, monkeys, apes and humans. Another solution to this taxonomic dilemma, which will be discussed in the following chapters, is to place the tarsiers in their own suborder, serving to recognize the similarities to the other suborders, as well as their own unique attributes (Figure 1.1).

Figure 1.1 Philippine Tarsier
(*Tarsius syrichta*).
(Photo: David Haring)

Overview of Primate Taxonomy (Table 1.1)

The **Order Primates** is subdivided into three **Suborders: Suborder Prosimii, Suborder Tarsiodea** and **Suborder Anthropoidea**.

Suborder Prosimii

The term Prosimii literally means "before (*pro*) monkeys (*simii*)," and indeed these primates lived in North America and Europe during the Eocene epoch some 53 million years ago, many millions of years before the appearance of monkeys. The suborder Prosimii is further divided into two Infraorders, Lemuriformes and Lorisiformes.

The tree shrews (Tupaiiformes), which were once included in the order of primates with the prosimians, presented primatologists with another taxonomic dilemma. These small, squirrel-like insectivores from South-east Asia have a number of primitive primate characteristics about the skull and teeth. However, it is now generally agreed that a number of their traits, such as the long snout, claws on all digits and multiple births, exclude them from classification as primates, and that the tree shrew represents a transitional form between the true primates and the early insectivores from which primates evolved. The importance of these rodent-like creatures is that, anatomically and behaviorally, they provide insight into the primitive primate condition (Napier and Napier, 1985). Nelson and Jurmain (1990:239) suggest that many experts have placed the tree shrew in a category all its own, the Order Scandentia "as a final act of taxonomic desperation."

Lemuriformes

Members of this infraorder, the lemurs, are found only on Madagascar, a large island off the southeast coast of Africa. The once numerous types of lemurs have been reduced to approximately 20 species, largely due to human interference in the forms of hunting and habitat destruction.

There are two groups, or Superfamilies, of Lemuriformes: the Lemuroidea and Daubentoniodea. The Lemuroidea include a number of species of varying sizes living in a wide diversity of ecological niches. The only **diurnal** prosimians belong to the Lemuroidea. In contrast, the tiny nocturnal aye-aye is the only extant representative of the Daubentonioidea.

Lorisiformes

The Lorisiformes are small nocturnal primates found in the forests of Africa and Asia. Two distinct forms, or subfamilies, are included in this classification of prosimians: the slow-moving lorises and pottos, and the nimble galagos.

Suborder Tarsioidea

The tarsiers, the only members of this suborder, are tiny nocturnal forest dwellers found only in southeast Asia. They exhibit a number of unique characteristics that will be discussed in detail in Part 2. However, their most dominant features are their enormous eyes, which have the distinction of being larger than their brains, as well as their stomachs (Fleagle, 1988).

Suborder Anthropoidea

This suborder, which includes monkeys, apes and humans, is divided into two Infraorders: the **Platyrrhini** and **Catarrhini**. These terms refer to the nasal characteristics of the primates in each group. The Platyrrhini, or New World monkeys, have flat noses with side-directed nostrils (*Platy* = flat, *rhini* = nose), whereas the Old World monkeys, apes and humans, with their downward-directed nostrils, belong to the Catarrhini (*Cata* = downward, *rhini* = nose).

These arboreal primates are distributed throughout the tropical regions of Central and South America, and from Central Mexico to Northern Argentina. The only Superfamily of New World monkeys, the Ceboidea, is divided into two families: the Callitrichidae and the Cebidae.

The Callitrichidae includes the marmosets and tamarins, tiny exotic looking creatures who live deep in the forests of South America, while the Cebidae are a much more diverse group of monkeys, both in physical size and type of social organization. They count among their numbers familiar species often found in zoos, such as the squirrel monkey, the capuchin (organ-grinder monkey), and the spider monkey.

Catarrhini

This Infraorder is divided into two Superfamilies, the Cercopithecoidea (the Old World monkeys) and the Hominoidea (apes and humans).

The Superfamily Cercopithecoidea is composed of one family of Old World monkeys, the Cercopithecidae, which is divided into two subfamilies: the Cercopithicinae and the Colobinae. The Old World monkeys are the most widely distributed group of nonhuman primates, and are found not only in the tropical forests and woodland savannas of Africa and Asia, but also in temperate regions such as the montaigne forests of Japan. The subfamily Cercopithecinae is the most diversified class of Old World monkeys, in terms of the number of species, absolute number of animals, body size, diet and ecological niche. In contrast, the subfamily Colobinae is a very specialized group of primates, many of whom subsist on a diet composed largely of leaves and are characterized by a sacculated stomach to aid in the digestion of cellulose.

Superfamily Hominoidea which includes apes and human, is composed of three families of primates: the Hylobatidae, the Pongidae and the Hominidae.

The Hylobatidae are the lesser apes, so-called because they are the smallest of the apes. The two types of lesser apes included in this family are the highly **arboreal** gibbons and siamangs (Genus *Hylobates*) of South Asia, who are noted for their ability to move quickly and gracefully through the trees.

The Pongidae are the largest of the nonhuman primates and, not surprisingly, are referred to as the great apes. Three genera are included in the Pongidae: the gorilla (*Gorilla*) and chimpanzee (*Pan*) found only in Africa, and the orangutan (*Pongo*) from the rain forests of Borneo and Sumatra.

The family Hominidae is composed of only one genus and species, *Homo sapiens* (wise man), so named by Linnaeus because of our ability to know ourselves, since "self knowledge is the first step on the road to wisdom" (Napier and Napier, 1985:13). Let us hope our legacy to the Order Primates, in terms of our stewardship of this planet, fulfills Linnaeus' predictions for the human species. A detailed discussion of the physical characteristics and life-way patterns of particular species will follow in Part 2.

The Primate Pattern

The success of primates, measured by the number of genera and species existing in the world today, lies in their generalized **morphology** and their behavioral flexibility (Napier and Napier, 1985). One implication of the unspecialized nature of primate anatomy is that there is no one feature that sets primates apart from other mammalian species. And of the 18 orders of **placental mammals**, only the primates cannot be identified on the basis of a specific characteristic (Stein and Rowe, 1989). Although this lack of specialization can be viewed as a primate trait, there is a constellation of evolutionary trends that characterize the primate order. W.E. Le Gros Clark (1959), a British Anatomist, first used the term "Primate Pattern" to refer to the general tendencies expressed to a greater or lesser degree by all primate species. This list of behavioral patterns, with some additions by Napier and Napier (1967), is still used to define the standard set of traits shared by primates.

The following general trends have been seen as progressive adaptations to an arboreal life-style which has characterized the primates from their first appearance in tropical forests some 65 million years ago (see Table 1.2).

1. Generalized Limb Structure

Primates have retained a relatively conservative physical form, while other mammalian species have become specialized. This lack of specialization has allowed primates the freedom to move about in a number of ways, using a variety of locomotor patterns. Primates have retained the **primitive** mammalian and reptilian pattern of **pentadactyly** (the possession of five digits on the hands and feet), which allows them to grasp branches as well as other environmental objects. Although ungulates, such as horses, have lost all but one digit and have evolved hooved appendages for speed of movement on the ground, traces of the pentadactylous condition can be seen in their embryos (Clark, 1962). The retention of the clavicle, or collar bone, is another primitive feature of primate anatomy that has been preserved. This provides more flexible shoulder joints than those seen in quadrapeds such as canines and felines, and confers the ability to rotate the arms and lift them above the head.

2. Nails Instead of Claws, and Tactile Digits

With the exceptions of marmosets and the aye-aye, primates have flat nails and fleshy pads at the end of at least two digits. These sensitive tactile pads confer a selective advantage to arboreal animals, allowing a more efficient contact with branches and a greater sensitivity to environmental surfaces.

3. Mobile Digits

Most primate species are able to hold objects with their mobile digits, while few non-primate mammals have this facility. The possession of an opposable thumb that can be rotated provides primates, with the exception of New World monkeys and prosimians,

Table 1.2 The Primate Pattern

1. *Generalized limb structure*—Lack of specialization in physical form and the retention of primitive physical traits such as five digits and the collar bone.
2. *Nails instead of claws, and tactile digits*—The possession of flat nails and fleshy pads at the end of at least two digits.
3. *Mobile digits*—The possession of manipulative digits and opposable thumbs which provides the grasping ability of primate hands and feet.
4. *Vision as the dominant sense*—The eyes are placed in the front of the skull, allowing for binocular and stereoscopic vision and a resulting elaboration of the visual areas of the brain.
5. *Shortened snout*—The reduction in the length of the snout, accompanied by a lessening of the reliance on the sense of smell and a reduction in the olfactory areas of the brain.
6. *Reduction in the number of teeth*—Heterodental dental pattern and generalized molars providing dietary flexability.
7. *Enlargement and Increasing Complexity of the Brain*—Larger brain relative to body size and greater elaboration of the cerebral cortex.
8. *Improved nourishment of the fetus*—Improved access of the fetus to the nutrients in the maternal blood.
9. *A prolonged period of infant dependancy*—All phases of development from gestation to adolescence are prolonged, resulting in an extended period of social learning.
10. *Single births*—Only one infant is produced at a time.
11. *A tendancy towards trunkal uprightness*—The trend for an erect upper torso, which culminates in bipedalism in humans.

with a **precision grip** and the ability to pick up very small objects. The manipulative digits on both hands and feet provide most species with a **power grip**. The grasping ability, or the prehensility of the primate appendages, is one of the primary adaptations for life in the trees (Cartmill, 1974).

4. Vision as the Dominant Sense

In modern primates, the eyes are placed in the front of the skull, allowing for an overlapping of the visual fields with the resulting **binocular vision** so important for depth perception. **Stereoscopic vision**, in which the hemispheres of the brain receive signals from both eyes, is another feature of primate vision. These attributes, plus the elaboration of **retinal cones,** which produce color vision, are critical for efficient movement in the trees. With the progressive increase in the reliance on vision as one moves from prosimians to monkeys to apes, the visual areas of the brain become larger and more highly developed.

5. Shortened Snout

The reduction in the length of the snout is accompanied by a reduction in the olfactory areas of the brain and less reliance on the sense of smell. The relatively flat face seen in most primate species also enhances binocular vision. The notable exception to the reduced snout is the baboon, whose **prognathic**, dog-like face is viewed as an adaptation to house the very large front teeth. **Terrestrial** mammals tend to have a more elaborate olfactory system, while a keen sense of smell is not as essential to an arboreal species (Clark, 1962).

6. Reduction in the Number of Teeth

Like most mammals, primates are heterodonts, meaning that they have various types of teeth that serve different functions. However, primates generally have fewer teeth than other mammalian species. The front teeth (incisors and canines) are useful for seizing and cutting food items, while the back teeth (premolars and molars) slice and grind the food to the point where it can be swallowed (Napier and Napier, 1985). The different types of teeth and the generalized, relatively simple structure of primate molars provide for the dietary flexibility needed for life in a variety of tropical and neo-tropical habitats.

7. Enlargement and Increasing Complexity of the Brain

Generally speaking, primates have larger brains relative to body size than do other mammals. There is also a difference in the organization of primate brains. The visual areas are more prominent, the olfactory center is smaller and the cerebral cortex is larger and more complex than in other mammalian species. There is a trend toward greater cerebral elaboration within the Primate Order, with the great apes having larger, mor convoluted neocortices than prosimians and monkeys. This expansion and elaboration of the cerebral cortex is reflected in primates' behavioral flexibility and enhanced ability to learn.

8. Improved Nourishment of the Developing Fetus

In most placental mammals, the nutrients from the mother's blood must pass through two sets of blood vessel walls to reach the infant's circulatory system. The infusion of maternal blood to the fetus is more efficient in most primate species, with the fetal blood vessels penetrating the wall of the uterus, allowing for the easy passage of nutrients (Stein and Rowe, 1989).

9. A Prolonged Period of Infant Dependancy

All phases of development, including gestation, suckling and adolescence, are prolonged in primates, providing an extended period to efficiently rear the young. Thus, generally speaking, more parental time and energy is invested in primate young that in other mammals of similar size, resulting in a greater dependancy on learned behavior. The time frame of all developmental phases and the entire life span increases within the Primate Order from prosimians to apes (Napier and Napier, 1985) (see Table 1.3).

Table 1.3 Duration of Life Periods of Primates

	Gestation (Days)	Infantile Phase (Years)	Juvenile Phase (Years)	Adult Phase (Years)	Life Span (Years)
Lemur	120–135	0.5	2	11+	14–15
Macaque	165	1.5	6+	20	27–28
Gibbon	210	2	6+	20+	30–40
Orangutan	264	3.5	7	30+	40–50
Chimpanzee	228	5	10	30	40–50
Gorilla	258	3	8–10	27+	40–50
Man	266	6	14	50+	70–75

Source: *The Natural History of the Primates* by J.R. Napier and P.H.Napier. Copyright 1985 by MIT Press. Reprinted by permission.

10. Single Births

Nonhuman primates, as a rule, produce only one offspring at a time, with the frequency of multiple births in most species being about the same as in humans. Further, time between births is prolonged, allowing for infants to be relatively mature physically and socially before the next one is born.

11. Tendancy towards Trunkal Uprightness

The trend for an erect upper torso in primates culminates in **bipedalism,** the ability to walk upright on two legs. This has been refined into a striding form in humans, and although almost all nonhuman primate species are able to walk on their hind limbs, they are not structurally suited to long periods of bipedalism. The upright positioning of the trunk also permits the free and precise movements of the upper limbs seen in all primates.

An important factor influencing the evolutionary divergence of the primates was an adaptation to an arboreal way of life. It can be clearly seen that most of these evolutionary trends are directed towards an enhanced ability to live in the trees. Primates found their adaptive **niche** in the trees, while other mammals adapted to a grassland or mixed environment (Relethford, 1989).

Locomotor Patterns

No other mammalian order shows such a wide diversity of locomotor patterns as the primates. Although all primate species have the same basic skeletal anatomy in terms of the number and types of bones, the structure of their limbs depends on the style of locomotion they normally use. The relationship between limb structure and method of locomotion has proven to be very useful in describing both living and extinct primates. A study of the limb structure of contemporary primates is particularly important to paleoanthropologists in the interpretation of the fossil remains of early primates, in that it helps them reconstruct the probable style of locomotion and to determine the ecological niche inhabited by the prehistoric form (Nelson and Jurmain, 1991).

Contemporary primates use the following four distinct forms of locomotion (see Table 1.4):

1. Vertical Clinging and Leaping

This is characterized by leaping in the trees, propelled by powerful hind limbs, and clinging to the target branch or trunk with the forelimbs. Vertical clingers hold their bodies erect while resting, grasping the tree trunks with their prehensile hands and feet. The upright trunk, elongated hind limbs and short forelimbs, which are the physical adaptations to this locomotor pattern, make hopping on the hindlimbs the only possible means of movement on the ground. **Vertical clinging and leaping** is only seen in the tarsiers and some species of prosimians, such as the sifaka, indri and galagos.

2. Quadrupedalism

Quadrupedal species generally hold their bodies in a horizontal position and move along the substrate using all four limbs equally. However, there are three main types, or sequences, of running and/or walking, depending on the speed of movement and the terrain (Napier and Napier, 1985).

   a. Slow Climbing and Walking - involves very cautious movement up the limbs of the trees and slow, deliberate walking along the branches. This gait is often associated with foraging for insects and is specific to the Lorisinae. These species never run or leap between branches, but depend on their hands and feet to pull themselves upward. Their hands, with widely opposable thumbs and shortened second digits, provide these small primates with an extremely stong power grip, allowing them to cling tenatiously to the branches.

   b. Branch Running and Walking - is the main mode of locomotion used by most New World monkeys and all arboreal Old World monkeys, allowing the animals to move quickly and easily along tree branches. These species leap from tree to tree, often using their tails as an extra limb to help them cross wide gaps in the foliage.

   c. Ground Running and Walking - is the predominant style of locomotion seen in **semi-terrestrial** species of prosimians and monkeys, such as baboons, vervets

15

Table 1.4 Nonhuman Primate Locomotor Patterns

Type	Description	Anatomical Adaptation	Example
1. Vertical clinging and leaping	Leaping in the trees and clinging with the forelimbs	Strong, elongated hind-limbs and tarsal bones, erect torso	Tarsier
2. Quadrupedalism	Using all four limbs	Limbs of equal length, flexible spine.	
a. Slow climbing and walking	Slow, cautious movement	Widely spread thumb, reduced 2nd digit	Slow loris
b. Branch running and walking	Moving quickly on branches and leaping	Limbs of equal length, flexible spine	Marmoset
c. Ground running and walking	Moving quickly on the ground	All limbs elongated	Baboon
3. Brachiation	Arm over arm swinging in the trees	Elongated forelimbs and hands, hook-like fingers, shortened thumbs	Gibbons
<u>Great Ape Locomotion</u>			
a. Modified brachiation	Arm over arm locomotion in the trees used faculta-tively	Elongated forelimbs and hands, hook-like fingers	Great Apes
b. Knuckle-walking	Terrestrial quadrupedalism, supported on knuckles	Long arms, short legs curved fingers.	Chimpanzee Gorilla
c. Quadrumanual Climbing	Slow movement from branch to branch using hands and feet	Flexible feet	Orangutan

Source: Adapted from Nelson and Jurmain 1991.

and macaques. Although these primates sleep in the trees and are adept at climbing and moving along the branches, they spend a large portion of their waking hours on the ground.

The major physical adaptations for all three styles of quadrupedalism are four limbs of approximately equal length and a flexible spine (Napier and Napier, 1985).

3. Brachiation

Brachiators move through the trees with the grace and agility of acrobats, using an arm over arm swinging motion. True brachiators seldom leap from branch to branch, but cover gaps in the foliage by using their bodies as pendulums to gain enough momentum to swing across. The specific anatomical requirements for this type of locomotion are long, strong arms, elongated hook-like fingers, shortened thumbs, short hind limbs and erect inflexible spines. Although the gibbons and siamangs are traditionally described as the only true brachiators, the New World spider monkeys, *Ateles* and *Brachyteles*, should also qualify as brachiators because of their consistent use of an arm over arm style of movement through the upper canopy of the trees and their typical brachiator limb structure.

The great apes could be described as modified brachiators because, although they are physically equipped to move arm over arm through the trees, they seldom do. Chimpanzees are semi-terrestrial primates who sleep in the trees, but spend about 50% of their waking hours on the ground and adult gorillas are tied to the ground because of their size and the lack of trees that can support them. Both of these species are able to walk bipedally for brief periods. However, their usual syle of terrestrial locomotion is called *knuckle-walking*, referring to their ability to move along the ground with their weight supported on the backs of their knuckles. The long arms, short legs and curved fingers, common to brachiators, facilitate this type of walking (Napier and Napier, 1985).

Orangutans, who are truely arboreal apes, deal with the problem of moving their bulky bodies through the trees by using a **quadrumanual** style of locomotion. This involves a slow and ponderous progression from branch to branch using both hands and feet to support the body. The feet of the orangutan are especially flexible and prehensile, which allows them to grasp branches easily, but make their version of knuckle walking rather difficult and ungainly (Napier and Napier, 1985).

Bipedalism

Although all modern primates are characterized by an erect trunk, and most species can walk upright on their hind legs for brief periods, only humans are able to stride and run upright efficiently (Napier and Napier, 1985). Many monkeys and apes could be referred to as **facultative** bipeds, since they only use this locomotor pattern for specific purposes, such as in displays to improve visibility. The human ability to use bipedalism as the habitual locomotor pattern has required a number of major physical adaptations, none of which are seen in nonhuman primates. The medium-sized body must be supported on inflexible feet with relatively short toes. The legs must be elongated, compared to the arms, and the pelvis must be altered for support to provide suitable muscle attachment areas.

The upper body must be totally erect, with the head placed squarely on a shortened and inflexible spine (Nelson and Jurmain, 1991).

It came as a surprise to field workers that all nonhuman primates species can use almost all types of locomotor patterns under certain conditions, which is additional proof of the benefits of the primate's generalized limb structure.

The Paleontology of the Primates

Considering that the age of the earth is estimated to be about 4.6 billion years, primates are relative newcomers to this planet. The first primate-like forms appeared approximately 65 million years ago, after the demise of the dinosaurs. A very quick overview of early life on earth will emphasize the important events that took place before the origin of the primates.

The last 570 million years for which there is evidence of life on earth has been divided into the following three eras of Geological Time, identified by the life-forms preserved in the rock:

The Paleozoic Era (Early Life) - 570 to 225 mya;
The Mesozoic Era (Middle Life) - 225 to 65 mya;
The Cenozoic Era (Recent Life) - from 65 mya to the present
(Napier and Napier, 1985).

Although the simplest forms of life came into being well before 570 mya, the beginning of the Paleozoic Era was marked by a rapid evolution of invertebrate species, marine forms that were very much like the jellyfish and sponges that we see today. This explosion of invertebrates was followed by the appearance, some 450 mya, of the first vertebrates in the form of jawless fishes, represented today by lamprey eels and hagfish. The later success of fish with jaws and teeth, and the evolution of amphibians from some of the many species of fish, paved the way for the movement of some of the Paleozoic amphibians onto land. These were to become reptiles capable of laying eggs with shells hard enough to survive on dry land. A radiation of these forms ushered in the Mesozoic Era, which has often been referred to as the Age of Reptiles. The dinosaurs, who were the most successful of the reptilian species, dominated animal life on the planet until the end of this era. Although mammals had their origins during the Mesozoic, about 200 million years ago, they were small, furtive animals living in the shadow of the overwhelming reptiles. The extinction of the dinosaurs 65 million years ago opened up a whole new world to the surviving mammals, who evolved a multiplicity of species to fill the vacant ecological niches. The species that adapted to life in the trees were the ancestors of the modern primates (Relethford, 1990).

An **adaptive radiation** is the rapid formation of a number of new species with different adaptive patterns that occurs when a population enters an area where there is little or no competition for resources. The Cenozoic Era, during which the primates evolved, has been divided into six Epochs, each of which is associated with one or more adaptive radiations of primate forms (Relethford, 1990) (see Table 1.5). Although there are now approximately 200 species of living primates, nearly twice as many fossil species have been discovered (Fleagle, 1988). Unfortunately, the fossil record of primate evolution is relatively sparse, with a number of gaps at critical time periods. However, there is enough material to give us a reasonably clear picture of our family tree.

Table 1.5 Adaptive radiations of nonhuman primates in the Cenozoic Era

Epoch	Millions of years	Adaptive Radiation
Paleocene	53–65	Archaic- or Proto-primates
Eocene	53–65	Prosimians
Oligocene	25–37	New World Monkeys
		Old World Monkeys
Miocene	5–25	Apes

The Paleocene Epoch (65–53 mya)

The first primate-like mammals, often termed **archaic-** or **proto-**primates, most likely evolved from the tree-dwelling, nocturnal insect-eaters of the Mesozoic Era. It is possible that the tree shrew, with its combination of primate and non-primate characteristics, is the nearest living example of a transitional form between these insectivores and the archaic primates (Napier and Napier, 1985).

One possible candidate for the earliest proto-primate is *Purgatorius*, which is represented in the fossil record by a single molar tooth dating to about 65 million years ago, found on Purgatorius Hill in Montana, U.S.A. Although the primate status of this tooth is not universally accepted, the bulbous cusp pattern on the molar is an indication of a diet of fruit and leaves. This represents a change from the pointed teeth commonly seen in insectivores, and could be the first indication of the dietary plasticity found in modern primates.

Judging by the wealth of fossil evidence for primate-like forms, the archaic- or proto-primates were firmly established by the middle of the Paleocene Epoch. By 55 mya, the Plesiadapidae family of archaic primates was the most abundant of the mammalian species in North America and Europe. These very diverse primate forms were small, squirrel-like animals with arms and legs that were well-adapted for climbing. The features of the skull were dominated by large protruding front teeth and prominent snouts. Although their eyes were still placed at the side of the face, there was evidence of an increasing dependency on vision. The **post-orbital bar**, the bony ring that surrounds the eye sockets, which is common to all modern primates, had not yet appeared. There was an indication of the ability to grasp, despite the presence of claws on all digits which suggests that, although they spent some time in the trees, they were not truly arboreal (Napier and Napier, 1985; and Relethford, 1990).

The widespread distribution of the Plesiadapidae in both North America and Europe may appear surprising, unless the effect of **continental drift** is taken into account. It was once thought that the earth's crust was solid and immovable. However, we now know that the continents have moved over millions of years, and do not occupy the same relative position today as they did in the past. During the Mesozoic Era, land masses were shifting, and the large northern and southern continents that had existed 180 mya were beginning to break up. By 55 mya, North America and Europe were joined at Greenland, and a body of water separated Europe from Asia. This allowed land mammals unrestricted access to North America and Europe, but not to Asia. Although the Plesiadapidae were a widely distributed and successful group in the late Paleocene and the early Eocene, they were later outcompeted by rodents and had become extinct by the middle of the epoch (Figure 1.2) (Napier and Napier, 1985).

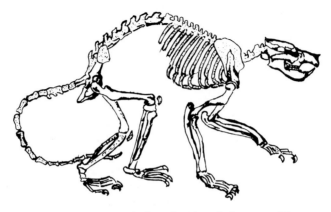

Figure 1.2 Drawing of Plesiadapidae skeleton ca 55 mya. Note the prominent snout, side-directed eyes and protruding incisors. (Redrawn from Napier and Napier 1985.)

The Eocene Epoch (53–37 mya)

The Eocene epoch, often referred to as the Age of Prosimians, is noted for the appearance of the first primates of modern aspect, who flourished in North America and Europe along with a great variety of other mammals. The family Adapidae, represented by the genera *Smilodectus* and *Notharcus* in North America and *Adapis* in Europe, was the most diverse and most well-represented primate form. The fossil remains of these two North American genera show a great number of anatomical similarities to modern lemurs. The Omomoyidae family, composed of at least 20 genera of early prosimians, were also very successful in North America. These small nocturnal animals, particularly members of the genus *Necrolemur*, with their large eyes and elongated tarsal bone, bore a striking resemblance to the modern tarsiers. However, it is not clear if these animals were ancestors of the tarsiers.

The Eocene primates were the first to exhibit the post-orbital bar common to all extant species. The eyes moved forward in the skull, nails had replaced some of the claws and the brain case grew larger, in relation to the size of the face. These characteristics all provided further evidence of the increasing adaptation to an arboreal way of life. By the end of the Eocene, tropical areas of the northern latitudes had shrunk, due to a general cooling of the land masses that had drifted away from the Equator. Rodents again appeared to outcompete the primates, and the close of this epoch seemed to signal the end of the prosimian dominance in all but the most isolated areas. Although a number of Eocene prosimians were extinct by the early Oligocene Epoch, it is speculated that at least some forms lived on and were the ancestors of modern prosimians. Perhaps it was during the Eocene that the ancestral lemurs rafted across the narrow Mozambique Channel from Africa on islands of vegetation to take up residence on Madagascar. It has been assumed that the anthropoids evolved from the Adapidae or Omomyidae, however no indisputable evidence has been found to support either form as our ancestor (Napier and Napier, 1985).

The Oligocene Epoch (37–25mya)

Fossil evidence, specifically an anthropoidal form found in Bolivia, points to the appearance of the New World monkeys in South America about 30mya. Although there had been a wide variety of mammalian species, such as ungulates and marsupials, living in South America in the Eocene and early Oligocene epochs, no fossil evidence of primates has been found in this time period. The burning question, still puzzling paleontologists today, is: Where did these ancestral New World monkeys come from? Since South America was an island at this point in geological time and completely isolated from North America and Africa, two possibilities for their origin have been suggested, and are still being vigorously debated (see Ciochon and Chiarelli, 1980 for a complete discussion):

1.   The Eocene prosimians from North America somehow drifted across the ocean separating the two continents and evolved into the monkey-like form. The problems associated with this theory include the complete absence of any physical characteristics common to both groups of animals, and the fact that the prevailing currents flowed from south to north, making any trans-oceanic journey from North America to South America very difficult.

2.   Monkey-like forms rafted over from Africa on tectonic islands that had broken away from the mainland. South America had separated from Africa, but the distance across the south Atlantic was much less then than it is now and the currents flowed in a westerly direction, favorable for the movement of small islands from Africa to South America. The morphological similarity between the Oligocene primates found in Africa and the South American forms tends to support this theory. However, the total lack of any Eocene primate fossil material in Africa or South America leaves the question still unanswered and, until more fossil evidence is found, the controversy is bound to continue.

Fossil finds indicate the appearance of Old World anthropoidal forms in the Fayum region of Egypt about the same time the New World monkeys appeared in South America (before 31 mya). This desert-like area, that was once a lush tropical rain forest intersected by slow-moving streams, has provided most of the fossil material for the Oligocene Old World primates. In contrast to the prosimians of the Eocene epoch, these forms showed a complex of anthropoid traits. The trend towards vision as the predominant sense was almost complete, with the eyes set fully in front of the skull and a greatly reduced snout. The smaller eyes indicated a transition to a diurnal life-style, which opened up a number of new opportunities, both in a practical sense, due to expanded food sources, and in the social dimension, because conspecifics must have become more visible. Although the Fayum primates were generally small, cat-sized animals, they were like modern Old World monkeys in many respects (Relethford, 1990).

One of the most interesting and relevant of the Fayum primates was *Aegyptopithecus* which was successful in the tropical forests of Egypt some 33mya. This fossil anthropoid bridges the gap between the Eocene prosimians and the Miocene apes, and represents one of the best transitional forms found in the primate fossil record. The dental characteristics were particularly significant, in that the number and types of teeth were typical of modern Old World monkeys, apes and humans, and the lower molars showed the **Y-5[1] cusp** pattern for the first time in evolutionary history. This cusp pattern in the mandibular molars is specific to hominoids, and is seen in no other primate form. The post-cranial skeletal remains were very monkey-like, with four limbs of equal length and evidence of a tail, prompting the description of *Aegyptopithecus* as "a monkey with apes' teeth." Not surprisingly, there is some speculation that this early transitional form is on, or near, the evolutionary line to the modern apes (Relethford, 1990).

The Miocene Epoch (25–5mya)

A major geological event that dramatically affected the evolution of life in the Old World took place in the Miocene epoch. North America and Europe had become separated in the Oligocene, and a land connection developed between Africa and Eurasia, allowing for the migration of species in both directions. There was a great uplifting of mountains that affected the climate and contributed to the shrinking of the tropical forests and the appearance of a mosaic of contrasting habitats, including woodland savannas, grasslands and semi-arid regions. These events set the stage for an adaptive radiation of Miocene primates who were distributed throughout Africa, Eurasia and China. Not surprisingly, this epoch is often called the "Age of the Hominoids" (Relethford, 1989).

These ape-like forms represented a very diverse group of animals, with a confusing number of genera and species that have been largely identified by their dental morphology. To avoid unnecessary confusion, only two genera which best typify the diversity of adaptations will be discussed.

Proconsul lived in the early and middle Miocene from 17 to 23 mya. From the fossil remains, it is clear that these primates first lived in the forested areas of Africa and later in Europe. Although the members of the genus were very diverse in size and other characteristics, most fossil forms showed a mixture of ape and monkey-like features. The dental morphology was very ape-like with the distinctive Y-5 pattern on the **mandibular** molars, and thin enamel and raised cusps on all of the molars, while evidence of four limbs of equal length was a more common monkey trait. The fossil remains indicate that *Proconsul* was an arboreal quadruped living primarily on a diet of fruit and leaves. It was very successful for millions of years, until the tropical forest began to shrink. The predominance of the ape-like characteristics of the fossilized teeth has prompted many paleontologists to classify *Proconsul* as a Hominoid.

Sivapithecus was a highly successful genus of Miocene primates from 17 to 8 mya. As with many of the other early primate forms, they were variable in size and geographical distribution, with fossil evidence in Africa, Asia and Europe. The massive jaws, holding large molars with thick enamel and low cusps, were the major distinguishing features of *Sivapithecus*. They indicated a change in diet from the soft fruit and leaves eaten by *Proconsul*, to one that included nuts, seeds and hard fruits. This would also seem to indicate that these late Miocene apes were not restricted to forested areas, but could live in a more open woodland savanna and grassland habitats. The **post-cranial** remains suggest that they were quadrupeds and not knuckle-walkers like the modern apes.

The exact place of *Proconsul* and *Sivapithecus* in the evolutionary scheme of things is uncertain, and the subject of a great deal of debate. Although one species could have been ancestral to the modern hominoids, the striking resemblance of *Sivapithecus* to the modern orangutan indicates that it is at least on the evolutionary line to the Asian ape. The Miocene fossil evidence is extremely varied and represents many different types—a number of which are quite unlike the modern apes. It is probable that several forms became extinct, and while we do not know which one was the common ancestor of the extant apes and humans, the family tree of the primates is clearly not a simple linear evolution, but more of a bush-like configuration with many branches and offshoots (Relethford, 1985).

---

[1]Mya refers to million years ago.
[2]"The Y-5 pattern refers to a pattern found on molars with five cusps separated by grooves reminiscent of the letter Y." (Stein and Rowe, 1989:492).

Chapter 2

An Historical View of Primate Studies

Although primatology is a relatively new discipline, the intellectual interest in nonhuman primates has been around for a very long time. The earliest record of monkeys dates back to about 1000 B.C., where there is evidence that Egyptians worshipped hamadryas baboons who were believed to have the ability to read and write. So great was the belief in their wisdom and virility that the God Thoth, who symbolized these attributes, was a male baboon. Worship was not restricted to the males, as the monthly menstrual periods of the female baboons were noted and believed to be associated with lunar cycles. The monkeys were kept in the temples, and the presence of their embalmed bodies with those of religious leaders is an indication of their elevated status in the early Egyptian religious system (Morris and Morris, 1968).

The Greeks were the first to attempt a scientific study of nonhuman primate anatomy. Their interest was piqued by the human-like appearance of the monkeys, and since the Greek scholars were forbidden to study anatomy by dissecting human cadavers, nonhuman primates were seen to be ideal substitutes. In 325 B.C., Aristotle wrote about the striking anatomical similarity between humans and monkeys. However, he believed in the Great Chain of Being, where every living organism was created as it presently existed, and so there was no suggestion of an evolutionary or biological relationship between the human and nonhuman primates. Hundreds of years later, the observations of Galen and Hippocrates, who also studied nonhuman primate anatomy, led them to agree with Aristotle's conclusions that the physical form of monkeys and humans was virtually identical (Richard, 1985).

Judging from the lack of literature on the subject, there was very little scholarly interest in nonhuman primates following the Greek fluorescence. The next mention of monkeys was in the Middle Ages, when thirteenth century Christians described them as fallen people who were in league with the devil. They emphasized the bestiality and disgusting behavior of the monkeys, and fostered a negative attitude towards nonhuman primates that endured for several centuries, which effectively discouraged scientific interest in them (Richard, 1985).

Attention was again directed towards monkeys in the 1500s, when explorers returned from their journeys to Africa, with stories of exotic creatures and living examples of monkeys and natives. About this time, Vesalius, a Belgian anatomist who had dissected both monkeys and humans, corrected misconceptions concerning the degree of anatomical similarity between the two, arguing that nonhuman primates were not identical to us in form and could not effectively take the place of human cadavers in anatomical studies.

Apes did not become known to the Western world until the seventeenth century when anatomists Nicholas Tulp and Edward Tyson conducted anatomical studies on chimpanzees (Richard, 1985).

In the eighteenth century, Linnaeus, a Swedish naturalist, published the first comprehensive classification of living plants and animals, and was the first to group humans, monkeys and apes together in the same category, that he named the Order of Primates. His classification, and the binomial system of naming plants and animals he developed, still forms the basis for modern taxonomic systems. Like Aristotle, Linnaeus believed in the constancy of all living forms, and based his classifications on similarities in appearance and anatomical characteristics, rather than on evolutionary or biological relationships (Napier and Napier, 1985).

It is interesting to note that, although the anatomical similarity between nonhuman primates and humans was often stressed and even exaggerated, in any discussion of behavioral similarity was discouraged. Unfortunately, in the eighteenth and nineteenth centuries, conceptions about the bestiality and savage nature of monkeys and apes were emphasized, based on the biased moralistic attitudes of the time, rather than on observational data. The external genitalia of the monkeys and apes were often very visible, and the animals in captivity tended to show a heightened level of sexual activity, all of which offended the sensibilities of early writers and scholars. Nonhuman primates were the favorite targets of cartoonists, almost always representing unpopular people or causes. Even eighteenth century scholars, such as Comte de Buffon, a French natural historian whose descriptions of nonhuman primates were generally insightful, could not hide his disgust for the appearance and sexual behavior of the baboon (Richard, 1985).

It was during the late eighteenth century that scholars, such as the philosophers Rousseau and Hoppius (a student of Linnaeus), began to realize that, although the anatomy of primates was well documented, little was known about their behavior. Although this deficiency was generally acknowledged, almost nothing was done to correct this situation until relatively recently (Reynolds, 1981). Monkeys and apes figured prominently in Darwin's evolutionary theories, and although his claims that they were our closest living relatives had gained general acceptance, the negative attitude towards their behavioral attributes persisted. It was permissible to discuss the anatomical similarities between human and nonhuman primates, but their behavior was not to be compared with ours. The most maligned species were the great apes, particularly the gorillas, who reportedly kidnapped young women, took them into the trees, and raped them (Morris and Morris, 1968).

Thanks to the investigations of twentieth century psychologists, scientific interest in nonhuman primate behavior was rekindled. The first truly scientific behavioral study was conducted almost by accident. Wolfgang Kohler, a German psychologist who was trapped in the Canary Islands during World War I, had access to a colony of juvenile chimpanzees. He disagreed wholeheartedly with the behaviorist tradition followed by American psychologists, and set out to observe the learning and problem-solving abilities of his ape subjects. He set up experiments to test the ability of the chimpanzees to reach inaccessible bananas by piling boxes on top of each other, or using sticks to rake in food items placed beyond their reach. Although the chimpanzees proved able to master the tasks he devised for them, Kohler's conclusions about the apes' ability to learn by a sudden flash of understanding were not totally correct, as the animals were familiar with the boxes and sticks before the tests began (Jolly, 1985). However, his insightful observations of chimpanzee behavior in his book <u>Mentality of the Apes</u> provided the academic world with the first truly scientific study of ape behavior and opened an important chapter in our understanding of nonhuman primates.

Robert Yerkes, an American psychologist, was struck by the amazing similarity between chimpanzees and humans, and became an important figure in the promotion of behavioral studies of apes. He and his wife collected all of the available knowledge concerning the major species of apes, and in 1929, they published <u>The Great Apes</u>, in which they stressed the pressing need for more information about these impressive creatures, particularly about their life in the wild. To date, there had been no really scientific field studies of primates, and all of the preconceptions about primate societies and behavior had come from zoos or captive colonies (Haraway, 1989).

One of Yerkes' most important contributions to primatology was the founding of the Laboratory of Primate Studies in Orange Park, Florida, under the aegis of Yale University which was one of the first major centers for the breeding and research of great apes. Yerkes was committed to the maintenance of detailed life-history records, and he tempered his scientific work on chimpanzee behavior with a real concern for their well being. Another equally important contribution to primatology was his training and encouragement of young scientists to go out to study primates in the wild. This was the real beginning of primate field studies. Harold Bingham, Henry Nissen and Clarence Carpenter were Yerkes' proteges who ventured into the field and studied primates with varying degrees of success (Haraway, 1989).

In 1929 and 1930, Bingham and Nissen traveled to Africa to study the social life and ecology of gorillas and chimpanzees, respectively. They took with them a large contingent of porters and guides who preceded them through the forests, scaring away any primates that might have been tempted to appear. The lack of contact with their primate subjects was partially responsible for the failure of their studies, and although they did gain some useful information, they were unable to provide any helpful hints for prospective researchers. It was 30 more years before Jane Goodall and George Schaller were to carry out successful field work on chimpanzees and gorillas (Southwick and Smith, 1986).

Yerkes' third protege, Clarence Carpenter, a post-doctoral student of comparative psychology, is generally acknowledged as the Founder of Primate Field Studies. In 1930, he began his landmark study of howler monkeys on Barro Colorado, an island in the Canal Zone, which provided primtologists with guidelines for the observation of free-ranging primates and methods of data collection which are still in use today. He was particularly interested in the dynamics of primate social groups, raising questions related to the formation and maintenance of these groups. His use of sociometric mapping to analyze primate social systems was an adaptation of a research technique from social psychology. In 1938, after completing a field trip to Thailand to study the social patterns of gibbons, Carpenter negotiated with the Indian Government for over 400 rhesus monkey to be shipped to Cayo Santiago, an uninhabited island off the coast of Puerto Rico. The monkeys were released and Carpenter observed them as they arranged themselves into social groups. He continued his research at Cayo Santiago until 1940, when he left to take a post at Pennsylvania State College. At this point, all research was suspended until after World War II, when scientists returned to the island and set up the Caribbean Primate Center, which continues to be a major source of data on rhesus monkeys (Haraway, 1989).

The publication of Carpenter's findings in the 1930s did not appear to stimulate any great surge in primate field work, and it was not until after the war that field research really "took off" (Southwick and Smith, 1986). The investigations of Carpenter and Dr.Solly Zuckerman, a South African anatomist, finally took root and triggered a great upsurge of interest in the behavior and social life of primates. Zuckerman, in his influential book The Social Life of Monkeys and Apes (1932), proposed that primate society was based on constant female receptivity, and that sexual bonds were the cement that held primate society together. Although his theories have been discarded, his book was important because it described primate society according to a set of principles, rather than in sentimental anthropomorphic terms (Morris and Morris, 1968).

The dramatic increase in interest in primate field work in the late 1950s can be attributed to a number of converging factors. The improved access to far-flung areas provided by jet travel, biomedical breakthroughs, the entry of zoologists into primate studies, and the revitalization of field research by primatologists such as Stuart Altmann, all served to stimulate activity in many areas of the world. Japan, India, Africa, the Caribbean and Latin America became the arenas for primatological research. Japanese scientists began to publish the results of field studies of their **indigenous** Japanese macaques in the early 1950s. However, it was not until 1957 that these were made available to Western scientists. The contribution of Japanese primatologists to our knowledge of primate social networks and concepts, such as kinship, should not be underestimated.

It is interesting to note that primatologists from different countries focused on very different aspects of primate biology and behavior. While the Japanese scientists were interested in the social dynamics of their macaques, cultural anthropologists from the United States were primarily interested in viewing nonhuman primate societies as models of early hominid social organization. In the early 1960s, American primatologists were conducting research in many tropical areas of the world. Phyllis Jay was studying the langurs in Northern India, while George Schaller dispelled the negative impression of the great apes by emphasizing the gentle nature of the mountain gorillas in Africa.

At the urging of Louis Leakey, National Geographic funded field studies of the great apes. Leakey's ability to recognize research potential was remarkable in his recruitment of Jane Goodall, Dian Fossey and Birute Galdikas to carry out the field work. These three women have since made an impressive contribution, not only to our knowledge of the life-way patterns of the great apes, but also to the conservation of these highly endangered primates. British field workers were interested in ecological studies, and directed their attention to the relationship between the environment and social behavior. Dr. K.R.L. Hall, an English psychologist, was one of the first scientists to recognize the influence of habitat on the social structure of primate societies (Richard, 1985). French primatologists, notably Jean Jacques Petter and Charles Domenique, specialized in studies of the prosimians, since Madagascar, the only place where lemurs exist naturally, was a French protectorate until 1960. Primate Behavior: Field Studies of Monkeys and Apes, edited by

Irven DeVore and published in 1965, was an important volume because it reflected not only the increase in the number of species that had been researched, but also the descriptive nature of the data (Richard, 1985). The scope of primate studies has widened over the decades to cover almost every aspect of primate behavior, biology and ecology. The research on the rhesus macaques of Cayo Santiago, Japanese macaque troops and the chimpanzees at Gombe, is particularly valuable to primatologists because it covers a number of decades and provides complete life histories of social groups and individuals (Southwick and Smith, 1986). The trend towards long-range studies on other species is underway, as we now realize the misleading and incomplete descriptions that can result from short-term studies, even though they may have involved thousands of hours of observation over a span of two or three years.

Although a New World monkey had been the subject of Carpenter's first study, the great majority of early investigations concentrated on Old World species in Africa and Asia. Baboons and macaques, being semi-terrestrial, open country forms, were studied because they were easy to observe, and the Great Apes were favored because of their likeness to humans. Less field research was directed towards New World monkeys primarily because of the difficulties involved in studying arboreal species. Of the ten genera that accounted for 60% of all publications from 1931 to 1981, only one (*Alouatta*) was a New World primate. Since the 1980s, this situation has changed, partly because of an increasing interest in all nonhuman primates, but also due to the pressing need to find out all there is to know about the severely endangered New World monkeys (Southwick and Smith, 1986). A similar situation existed for prosimians, who were nocturnal and arboreal, and therefore doubly difficult to study. Fortunately, long-term projects are being established, at least on the lemurs of Madagascar, and the endangered status of many of the species is the object of conservation efforts.

The decline in the world-wide populations of primates in the wild is now a major concern which has added a sense of urgency to field research. Over 50% of primate species are in some sort of jeopardy, primarily due to the disappearance of tropical forests. However, there is evidence that primatologists can make a difference to the survival of wild populations (Southwick and Smith, 1986). The publicity surrounding the work and untimely death of Dian Fossey at Karisoke Research Station in Rwanda has focused international attention on the plight of the mountain gorilla. Primatologists continuing Fossey's research have habituated at least two groups of gorillas to human presence and are allowing tourists to view them for only one hour in their natural habitat. This popular tourist attraction has added much-needed money to the Rwandan economy, and has raised the awareness of the local people of the gorilla as a valuable natural resource that must be protected. In South America, the efforts of primatologists have not only been directed towards introducing captive-bred golden lion tamarins into the forests of Brazil, but also towards educating the people about the importance of protecting their monkeys. If primatologists can save only a few species in the wild, their efforts will have been worthwhile. If the worst possible scenario should come to pass and some species disappear from their natural habitat, at least we will have the knowledge of their physical and social needs in order to successfully breed them in captivity.

Chapter 3

Primate Ecology

Introduction

There are a number of definitions of **ecology**, some more detailed than others. However, they all basically refer to the relationship between the individual organism and the environment. Early ecological studies were directed at the group level, attempting to find explanations for social systems by comparing the habitats exploited by different primate groups. The investigation of the relationship between the environment and social organization really began with Carpenter's field work in the 1930s. He established a research tradition for ecological fieldwork that dominated subsequent studies for many decades, whereby diet and habitat use were described in meticulous detail with little attention given to theoretical issues. Crook and Gartlan (1966) were the first primatologists to formalize the adaptive relationship between primate social organization and the environment by proposing a classification of primate species based on feeding behaviors, activity patterns and habitat use (Richard, 1985). Their work stimulated further efforts in this area and, although some relationships between the environment and social patterns have been found, the evidence is not strong enough to be of real predictive value (Rodman and Cant, 1984).

In the 1970s, two separate academic developments combined to precipitate a shift in ecological thinking to an emphasis that ecological factors could affect the behavior of individual group members. The first stimulus was the introduction of the new discipline of sociobiology, whose mandate to establish a biological basis for social behavior was promoted by stressing "fitness" in terms of reproductive success. This stimulated interest in the ultimate, indirect evolutionary factors that affect behavior, and generated important questions about the adaptive function of behavior. Secondly, the publication of Jeanne Altmann's paper (1974) "Observational study: sampling methods" in the journal Behaviour, led to improvements in the methods of observation and data analysis. The **focal animal** sampling method, which provided so much information on the behavior of the subject animal, became a popular observational technique which also directed the attention of scientists to the individual. The study of the relationship between social systems and the environment has expanded to include the individuals making up the systems (Rodman and Cant, 1984).

Ecological studies of primates present difficulties not encountered in similar research conducted on other social mammals who rely on a diet of plant foods. Predictive models are more difficult to establish for primates because of their **dietary plasticity** and behavioral complexity. Whereas many mammals depend on plant foods that are relatively constant in nutritional value, varying only in availability, many primates are able to eat a number of plant parts (flowers, leaves, fruit, grasses, seeds, etc.) which vary widely in nutritional quality. Since primates represent the majority of the arboreal **herbivores** living in the tropical forest, the ecology of the primates has proven to be particularly valuable in our understanding of the ecology of the rapidly disappearing tropical forests (Richard, 1985).

Primate Distribution

Primates have lived in the tropics since their appearance on this planet some 55 million years ago. Today, there are approximately 200 species living in all the tropical **biomes**, including rain forests, woodlands and savannah mosaics and the semi-arid brushland of Africa, Asian, and Central and South America (see Figure 3.1). Primates, other than humans, have never lived in Australia, New Zealand, New Guinea or Antarctica. This is likely due to the presence of a deep sea that existed before the appearance of primates, and prevented their movement from Southeast Asia to these areas. Although primates are most numerous and diverse in the tropical forest, they are widely distributed in all areas of the tropics and subtropics, and representatives of the two subfamilies (Cercopithecinae and the Colobinae) of Old World monkeys live in temperate zones in Asia and Africa. The rhesus and stump-tailed macaques (*Macaca mulatta* and

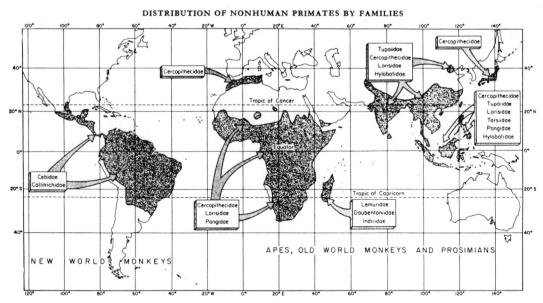

Figure 3.1   The distribution of nonhuman primates by families. Stippled areas show the approximate ranges of nonhuman primates. From *A Handbook of Living Primates* by J.R. Napier and P.H. Napier. Copyright 1967 by Academic Press Inc. (London) Ltd. Reprinted by permission.

Macaca arctoides), the Hanuman langur (*Presbytis entellus*) and the golden monkey (*Rhinopithecus roxelanae*) live in the eastern and southern slopes of the Himalayan foothills, while the Barbary macaques (*Macaca sylvanus*) inhabit the Atlas mountains of Morocco. The Japanese macaques (*Macaca fuscata*), the most northerly living nonhuman primates, range throughout the snowy mountain slopes of Japan (Richard, 1985).

These species are living examples of the behavioral flexibility seen in primates because they do not have any of the physical adaptations that allow other similar-sized mammals to cope with the cold winters of temperate zones. They do not hibernate or sink into a torpor like ground squirrels or bears, nor do they have appendages adapted for digging under the snow for the best quality food. They do not have the habit of storing food like squirrels, and lack the specialized teeth of the beaver and rabbit useful for a diet of browse. Although the langurs and golden monkeys have a large rumen-like stomach, allowing them to digest fibrous foods, it is unlikely that this adaptation to leaf-eating is very helpful, given the low nutrient content of their winter diet. The thick winter coat grown by the macaques species is the only physical accommodation to life in a cool climate.

It appears that only their behavioral adaptations allow these species to survive during the cold winter seasons. All are semi-terrestrial, spending a large proportion of their waking hours on the ground searching for food, since most of the nutritious food is found on or under the snow. The Japanese macaques change their eating habits drastically, subsisting on bark, needles and cones in areas of heavy snowfall, and seeds, grasses and roots where the snow cover is light. The langurs are able to exploit grassy meadows in some areas of Nepal, and subsist on mature leaves, roots and crop remnants in other parts of the Himalayan foothills. The monkeys tend to frequent the sunnier southern slopes of the mountains and huddle for warmth during storms and at night. In some areas of Japan, natural hot pools provide some respite from the cold for the macaques (Richard, 1985).

Few medium-sized herbivores can exist outside of the tropics because of the difficulty of extracting nutrients from low-quality vegetable food. Although nonhuman primates have the added problem of competing for high-quality food when they are not physically adapted to compete, it is possible that the ability of some species to live in temperate regions is related to human occupation (Richard, 1985). The food provided by Japanese primatologists, and the people of India who look upon the rhesus monkeys and the langurs as sacred animals, as well as the availability of crop remains likely help some of the primates over the lean winters.

Small primates, such as the arboreal marmosets and tamarins, who must live in the trees, would be unable to survive in temperate regions where high quality food items are found primarily on the ground. They lack the physical adaptations, such as the ability to hibernate, that allow other small mammals to survive in winter. At the other extreme, the great apes need substantial quantities of nutritious food, not only to provide the energy for survival, but also to fuel their large, complex, energy-expensive brains. Although none of these large primates have physical specializations allowing them to digest a diet of browse, it is doubtful that they could survive on such nutrient-poor food even had they the necessary adaptations (Richard, 1985).

The modern primates, regardless of size, with their powers of visual discrimination, their hand-eye coordination and their grasping hands and feet, are well-equipped to exploit the resources of the tropical forest. In fact, their distribution is dictated by the need for the readily accessible supply of high-quality food provided by the tropical forests, and this is where the primates are most numerous and diverse (Richard, 1985). However, the rain forest is a complex ecosystem, and taking advantage of the largesse available requires a wide knowledge of the environment. Although the food supply in the tropical forest may appear to be boundless, the incredible diversity of plant forms and the variability in their patterns of leafing, flowering and fruiting makes successful foraging a complicated process. There are predictable seasonal peaks and valleys in the abundance of food items; however, unforseen circumstances, such as droughts or torrential rains, could upset the system. The presence of tough skins, husks or toxins which affect the edibility of a plant species is an added constraint on foraging success (Oates, 1987). Given that the brain is an extremely energy-expensive organ to operate, there would have been selection for a reduction of brain size had not the large brains of primates provided an adaptive advantage. It is possible that the expansion of the neo-cortex in primate brains allowed them to become specialists in securing high quality, readily digestible food required to fuel their large brains (Richard, 1985). Richard (1985:121) suggested that "Being a finicky eater with a big brain carries a host of implications, not only for a primate's foraging behavior, but also for its patterns of reproduction and social life."

The tropical forest, which is the home of the majority of primate species, actually covers a wide range of habitats which vary in climate, topography, soil types and vegetation. The primary rain forest is characterized by a continuous canopy of tall trees and a dark understory devoid of any vegetation except for vines and tree trunks. In the secondary forest, where the canopy is much lower, more light is available, allowing for an abundance of leaves and fruit. Woodlands, consisting of short deciduous trees interspersed with low bushes and gallery forests lining the rivers are also home to a number of primate species. These different habitats present the primates living there with different sets of resources, such as food, water and safe sleeping sites, all of which require different adaptations. Within each of these habitats, particularly the primary forest, species are able to live sympatrically because they exploit different levels of the forest canopy and utilize a wide variety of food items (Fleagle, 1988).

In terms of diet, the general rule for primates is plasticity or eclecticism, with most species able to eat a variety of food items which include fruit, flowers, leaves and some insect or animal matter. However, often a particular species or group of species will have a more specialized diet and eat a preponderance of one type of food, allowing primatologists to classify them in one of three broad dietary categories: **frugivores**, **folivores** or **omnivores**. The term frugivore refers to those primates species subsisting mainly on fruit, while folivores depend on a diet of leaves to sustain them. Omnivorous species eat a wide variety of food items. It should be pointed out, however, that these are gross categories, and most primates show a degree of dietary plasticity, allowing them to substitute other food items when their preferred foods are scarce or unavailable (Oates, 1987).

The dietary needs of a species have implications for its daily activity patterns. Given that edible foods are often patchily distributed in the tropical forest, and that leaves are far more abundant than fruit, folivores are not required to travel as far for their food as frugivores. However, since leaves are harder to assimilate, folivores sit around for hours waiting for their food to digest, and tend to be less active than fruit eaters (Oates, 1987).

Habitat Use

In order to deal with the complexities of their environment, primate groups tend to restrict their activities to a specific area of the forest that they know best (see Figure 3.2). This is the groups' **home range**, which refers the the total area utilized over an extended period of time, while **day range** describes the distance an individual or a group travels during a day (or night). Although the home range may cover a relatively large area, there is usually one particular part of the range that is used more extensively than the rest because it contains the richest food sources and/or the safest sleeping sites. This is the **core area**, where much of the social activity takes place and consequently is often the part of the range that is most interesting to the primatologist (Napier and Napier, 1985). Frequently, the home ranges of a number of groups of **conspecifics** (members of the same species) will overlap, precipitating vocalizations and displays, The area of the home range that is actively defended against interlopers is referred to as the **territory.**

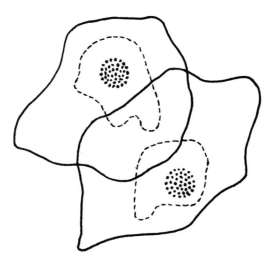

Figure 3.2   Home Range (__), daily range (_) and core area (. . . . .) of two hypothetical primate groups.

Territorial defense is a very labile characteristic, depending greatly on the species and the distribution of food. Research has shown that, on the whole, primates are not particularly territorial. Some species show territorial defense only on occasion, while some, like the gorillas, have a particularly *laissez-faire* attitude to overlapping groups. While primates that live in monogamous family groups tend to be very territorial, gibbons are the only primates that consistently and regularly defend their fruiting trees (Jolly, 1985).

The size of home ranges varies widely between species depending primarily on resource availability. Primates tend to have large home ranges in dry areas where resources are scattered, however in forested areas where the resources are distributed vertically as well as horizontally, the home ranges may be relatively small. For example, the daily range of a baboon on the African savannah can be as extensive as 19 kilometers, while the slender loris may only travel 30 meters daily in the forests of Sri Lanka (Napier and Napier, 1985).

Ecological Principles

The ecological niche of a species refers to the habitat that precisely fits its requirements for life in their environment. Although the potential niche for a particular species is bounded by a number of factors, such as climate, spatial needs, food availability and predation, the actual niche is dependent on the degree of interspecific competition for resources. Many arboreal tropical forest-dwelling primates live in mixed groups, where other species form an important component in their daily lives. At first glance, one might be led to believe that primates have been able to defy the principle of competitive exclusion, and coexist with complete competitors. **Complete competitors** are defined by Jolly (1985:35) as "Species governed by the same limiting factors in a community at equilibrium." In reality, **sympatric** species of primates, who occupy the same general ecological niche, are separated by one or more factors. They may be spatially separated and occupy different levels of the forest canopy, they may have slightly different dietary needs, or they may be active at different times (Jolly, 1985).

These mixed groups of primates may be chance associations of species who are attracted to the same food source, or they may be more stable unions of species that are attracted to each other for some reason. Traveling in mixed groups can have a positive or negative influence on the life-style of the respective species. One clear indicator of competition in a species is population density, where the number of animals in a social group increases when competition is absent (Waser, 1987). From census data on primates in the Ugandan forests, it is suspected that the density of *Cercopithecus mitis* is higher when competitors are absent than when surveyed in mixed groups with other species (Struhsaker, 1975). Extreme competition may act to eliminate a species from a particular area, or escalate the intensity of territorial behavior between species. Shifts in diet and ecological niches may also provide another indicator of the effects of competition, particularly during seasons when food is scarce. Field studies in Gabon, Peru and Malaysia have shown that the diets of species living in polyspecific groups overlap when resouces are abundant and are most divergent during the dry season (Waser, 1987).

Although it might be predicted that primates would forage for food more widely when living with other species than when competition was absent, this does not seem to be the case. Research has shown that there is little difference in the rate of movement of species when traveling in **polyspecific groups** than when traveling alone. The explanation for this is likely related to the fact that the impact these associations are not always negative. Squirrel monkeys (*Saimiri*), who often forage in polyspecific groups with the larger cebus monkeys (*Cebus*), benefit from this association by scavenging food from hard-shelled nuts that are inaccessible to them, and also by the improved ability to locate food items (Terborgh, 1983). Since the cebus monkeys appear to gain nothing from these associations, Terborgh (1983:176) has suggested that they are ". . . mildly parasitized by the *Saimiri*." Semi-terrestrial species, like baboons, often travel under groups of arboreal monkeys to gain easy access to the fruit that they drop.

Added protection from predators could be viewed as another advantage of a mixed-species association. Although large groups of primates could be easily located by predators, the benefits in terms of early detection of raptors and other prey species and safety in numbers would surely outweigh the risks. Since hybrids, resulting from the cross-breeding of different *Cercopithecus* species are often viable, there may be the social benefits of access to grooming and reproductive partners for solitary males in mixed primate groups. The incidence of stable associations between certain species, such as *Saimiri* and *Cebus* in The New World and *Cercopithecus mitis* and and *Cercopithecus ascanius* in Africa, indicates that there could have been selection for mixed-group associations in primates, either for added efficiency in foraging or for protection against predators. The evidence that competition has selected for niche differences between species is less clear (Waser, 1987).

Arboreal primates living sympatrically in the tropical forests must be specialists in their adaptations to the environment in order to coexist with other species with similar ecological requirements. There is a continuum between the narrow niches of the arboreal specialists and the large niches of highly adaptable species, like baboons and macaques, who use a wide diversity of environments. The baboons are generalists, who can live in the woodland savannah and open grasslands, but are outcompeted in the dense tropical forest by the forest dwelling *Cercopithecus* monkeys (Jolly, 1985).

R and K Selected Species

Primates are among the mammalian forms who have been defined as K selected species, because of their adaptations for survival in a relatively stable environment which is near its carrying capacity. K selected species have evolved a reproductive strategy in which the few offspring produced are provided with relatively long periods of parental care in order to ensure that their survival in a competitive environment. Large bodies and prolonged life cycles are characteristic of K selected species. In contrast, r selected species, who live in unpredictable environments where major perturbations frequently wipe out a significant proportion of the population, are selected to produce large numbers of offspring and spend little or no time in caring for them. Although many of the young will perish from environmental agencies, some will survive to reproduce. These species are small-bodied, with shortened gestation periods, a limited duration of immaturity and brief life spans (Jolly, 1985).

The distinction between r and K selected species is relative, with fish who produce millions of eggs at one extreme, and large-bodied, slowly reproducing mammals, such as apes and elephants at the other. Even within the primate order, there can be a significant variation in the degree of K selection between groups of a particular species, depending on the stability of the environment.

Summary

The study of primate ecology was once devoted to the meticulous description of dietary and overall adaptations to the habitat in an attempt to find correlations between the environment and the type of social organization. The scope of ecological research, in terms of theoretical orientation, has expanded considerably in the last decades and is now focused on the effects of the environment on the behavior, biology and demography of individuals and social groups.

The inability to find a significant relationship between the environment and the primate social organization is not surprising, given the dietary plasticity and behavioral flexibility characteristic of primates in general. Although the majority of primate species are restricted to life in the tropical biomes of Africa, Asia and Central and South America, the impressive degree of diversity in their diets and patterns of habitat use is rarely found in other mammalian orders.

The size of the daily range, the overall home range of primates, and the population density of primates is often dictated by the availability of resources (food, water and safe sleeping sites) and the degree of competition for these resources from other sympatric species of primates. The number of primates that live in mixed-species groups in the dense tropical forests indicates that small differences in diet and spatial and temporal adaptations produces ecological specialists. The generalists, like baboons, who range widely over a variety of environments, are outcompeted in the tropical forests by these specialists within their small niches.

Since primates live in the relatively stable environment of the tropical forest, they have evolved as K selected species. Their few, carefully reared offspring are well equipped to survive the expected competition for resources in a habitat nearing its carrying capacity. Jolly (1985:35) summed up the distinction between the K selected primates, and the rapidly-reproducing R selected species, when she wrote "Mental agility has in part replaced reproductive agility."

Chapter 4

Social Organization

Introduction

All primates live in some sort of social grouping which endures over time. Some species live in tightly knit groups that are easy to identify, while others live in fluid groups that coalesce and disperse into smaller units, making identification difficult. Some primates live in small family units, whereas others form very large troops consisting of hundreds of animals. Clearly, there is an enormous variation in the types of social groups exhibited by different species, and sometimes there is even variation in the makeup of the social group within a single species (Fedigan, 1992).

A **social group** is composed of animals that interact on a regular basis. These primates are able to recognize each other and spend more time in the company of group members than with other conspecifics. The term **social structure** refers to the physical makeup of the group in terms of age, sex classes and their relationships to each other. **Social organization** is a much more inclusive expression generally used to describe several aspects of the social groups, including spatial distribution, group composition and the social and physical relationships within the group. The primary difference between social structure and social organization is that the latter includes a behavioral component (Richard, 1985).

It is the nature of scientists in all disciplines, including primatology, to clarify their thinking by categorizing aspects of their study, such as species or social groupings, Primatologists have traditionally attempted to classify primate social groups on the basis of ecological factors, such as habitat use and diet, or in terms of socio-sexual behavior based on mating patterns.

Socioecological Classification Systems

A number of classification systems were developed in an effort to find relationships between environmental agencies and social organization. Crook and Gartlan (1966) were the first to attempt such a classification by dividing primates into five adaptive grades based on social structure, diet, habitat, and activity cycle. They suggested that their system represented an adaptive progression in which environmental factors selected for specific physical and social attributes (Table 4.1). This early work was important in that it stimulated an interest in **socioecology**, and their efforts were followed by others who attempted to improve on their scheme (Aldrich-Blake, 1970; Crook, 1970; Jolly, 1972; Eisenberg et al., 1972; Altmann, 1974; Clutton-Brock & Harvey, 1977; Richard, 1981). However, correlations between aspects of primate social organization and ecological factors proved to be inconsistent, and could not be used with certainty for all species. The basic problem with these ecological classification systems was the ambiguity of the ecological and social factors that the primatologists were attempting to define. It is very difficult to be precise when defining patterns of social behaviors and diet to be used as variables in statistical correlations. Questions such as "How much fruit does a species have to eat to be classified as a frugivore?", or "How much time does a species have to spend in the trees to be classified as arboreal?" present real problems. The choice of variables used in the environmental correlations were often arbitrary, ranging from physical factors such as body size and the degree of **sexual dimorphism**, to behavioral attributes such as dominance and territoriality (Richard, 1985).

A comparison of the forest-dwelling red colobus monkey(*Colobus badius*) and the black and white colobus monkey(*Colobus guereza*) (Struhsaker and Oates, 1975) provides a classic example of the problems encountered when attempting to use gross classifications of habitat and diet as indicators of behavior and social organization. These closely related species of monkeys are classed as arboreal folivores and both have the sacculated stomach which is specially adapted for digesting fibrous material. However, they exhibit a number of differences in their social patterns. The red colobus monkeys live in large, multi-male, multi-female groups whose extensive home ranges overlap with groups of conspecifics and other species without any overt displays of territoriality. They are relatively silent animals, who emit only a few low-intensity vocalizations. In contrast, the black and white colobus monkeys live in small heterosexual groups with one or two males, some females and their infants. They typically defend their small home ranges with loud, booming calls. Although both species subsist on a diet made up primarily of leaves, a closer look at their respective food preferences could shed some light on the differences in their life-styles. The varied diet of mature and immature leaves, buds, shoots and fruit from a number of tree species consumed by the red colobus monkeys requires them to range more widely than the black and white colobus monkeys who eat a monotonous diet made up primarily of the leafy products of only one common species of tree (*Celtis durandii*). Struhsaker and Oates (1975) suggested that the small, cohesive groups and narrow home ranges of the black and white colobus are adaptations to the efficient exploitation of a high density, uniformly distributed food source, while the more varied fare of the red colobus requires greater inter-individual spacing and perhaps a larger home range. The territoriality displayed by the black and white colobus could be also be a function of their restricted diet and small ranges. Therefore, despite similarities in general morphology, habitat and diet, these two species show important differences in social organization. This example illustrates the problems associated with making broad statements about the correlations between diet, home range and group size, when in fact the relationships are not as simple as they may seem.

Socio-sexual Classification Systems

The socio-sexual categorization of social organizations is based on the mating patterns exhibited by the primate species. Mating patterns on nonhuman primates fall into two broad types: **monogamy** and **polygamy**.

1. Monogamy is a pattern that describes a mating pair: a male and a female that breed exclusively with each other.

2. Polygamy is a pattern in which an individual mates with more than one member of the opposite sex. There are three variations of polygamous social groups that can be found in primate societies: (a) **Polygyny,** where one male mates with several females; (b) **Polyandry,** where one female mates with more than one male; and (c) Multi-male, multi-female groups, where several males and females live and mate together.

These mating patterns provide a convenient way of classifying primate societies without presupposing any relationships between the ecological, physical or social aspects of primate life. Monogamous primate groups are fairly rare, being restricted to a few species in each major taxonomic group, while polygynous and multi-male groups are relatively common. Nocturnal prosimians often live in a variation of the polygynous society where the large range of the adult male overlaps with the smaller ranges of several females. Thus, although the male does not actually live in the same social unit as the females and their infants, he interacts with the females often enough to mate with them. The only example of polyandry in primates that has been found are some groups of saddle-back tamarin where there is evidence of two or three adult males sharing a breeding female (Terborgh and Godizen, 1985).

Monogamous Societies

This type of nonhuman primate social group, composed of one male, one female and their offspring, is often called a family group because it resembles our typical notion of a Western society family. Monogamous groups are relatively uncommon in primates, restricted to only one species in each of the taxonomic groups of tarsiers, prosimians and Old World monkeys, but is widespread in the families Hylobatidae, Callitrichidae and Cebidae. Although the monogamous species are not closely related phylogenetically, they appear to have the following characteristics in common (Fedigan, 1992):

1. Lack of sexual dimorphism - The male and the female are the same size and often have the same coloration.

2. Lack of differentiation in sex roles - The male and female participate equally in protection against predators and territorial displays against conspecifics.

3. Territoriality - Monogamous species tend to be extremely active in the defense of their territories against conspecifics.

4. Extensive paternal care - The male invests a great deal of time and energy in the care and raising of the offspring.

5.   Synchrony of activities - The male and female forage, feed and rest at the same time.

In some species, particularly the marmosets and tamarins, there appears to be a delay in the sexual maturation of the young while they remain in the family group.

Polygynous Societies

This uni-male, multi-female social grouping is widespread among primate species in all ecological habitats, ranging from open country areas of Africa to the tropical forests of Africa and Asia. Although there is considerable variation in habitat use and the expression of social behaviors, most polygynous groups have the following characteristics in common:

1.   Closely-bonded females - The females are very cohesive and supportive of each other, and there is evidence, at least in some polygynous societies, that the females may be related.

2.   Peripheral male - The male is often peripheral to the main social group and does not have a central role in group activities.

3.   All-male groups - In addition to the heterosexual groups, there are always units made up entirely of closely-bonded males.

4.   Intolerant males - The male in the heterosexual groups is very intolerant of other sexually mature males.

5.   Male tenure is not permanent - The position of the male in the heterosexual groups in not permanent, as his tenure is often challenged by solitary males or males from the all-male groups. (Fedigan, 1992).

The solitary life-style of nocturnal prosimians, where the range of an adult male overlaps the smaller ranges of several females, is a variant of this social pattern. The basic difference is that the male does not live in the social unit with the females and her offspring.

Multi-male, Multi-female societies

This type of social group, where a number of adult males and adult females live together with their offspring, is common in primates but relatively rare in other animal species. The ratio of males to females is usually about 1 to 2, which is likely a function of a pattern of male emigration and female **philopatry** exhibited by a number of these societies (Napier and Napier, 1985). In these groups, the males migrate towards the periphery of the group when they reach puberty, after which they may leave and lead a solitary existence or join another group. The females remain in their natal groups, living out their lives in close proximity to their female kin.

The distinction between a uni-male and a multi-male society is often blurred for, as Eisenberg et. al. (1972) suggested. In some cases, such as the gorilla, there may be only one breeding male living with the females and his semi-adult and juvenile sons.

Any system of classification is subject to problematic forms that do not seem to fit any category; nevertheless, some frame of reference is needed to facilitate the description and the comparison of primate social groups. Since the classification of primate societies in terms of their mating patterns is less ambiguous and therefore more convenient than using ecological factors, it will be the system of choice for the following chapters.

The Evolution of Social Groups

Before field studies of nonhuman primates were common, scientists speculated that all primate groups lived in polygynous groups with one large dominant male and several females. As it happened, the first species to be studied in the wild were the baboon, the gorilla and the common langur, all of whom lived in multi-male, multi-female groups. So, it was argued that possibly this was the original social group from which all the others evolved. The large multi-male groups would break down to polygynous societies in areas where resources were scarce, since a sparse environment would only be able to support one large male and a number of smaller females, who would obviously require less food (Fedigan, 1992).

Eisenberg and others (1972) argued that the opposite scenario was likely closer to the truth, given the proportionately large number of uni-male groups compared to multi-male groups. Their model for the evolution of primate societies was based on the increasing tolerance of adult males, first towards adult females, then towards other adult males. The model suggested that the original primate society was that of nocturnal prosimians, with the range of the males overlapping that of several females and their offspring. If the males became permanently attached to one of these matrifocal units, the resulting social system would be a monogamous family group, with one mating pair and their infants. However, if the male became bonded to a number of the females within his range, a polygynous society would be the result. Eisenberg et. al. (1972) introduced the term **age-graded society** to describe a society in which the male became tolerant of his semi-adult sons and allowed them to remain with the group. Finally, in some cases, the male could become tolerant of other unrelated males, giving rise to bisexual groups consisting of a number of males, females and their infants.

Why Primates Live in Social Groups

It is clear that primates are particularly social animals and, although some species appear to live relatively solitary existences, all known species have been seen to exhibit complex systems of communication, ranging from olfactory signals, to gestures, to vocalizations. Given that primates have proven to be successful over evolutionary time, their social way of life must have conferred upon them a selective advantage that has allowed them to survive for approximately 55 million years.

It has been suggested that social living would give the members of groups advantages in terms of access to food, protection from predators, access to mates, and assistance in rearing offspring (Fleagle, 1988). However, it is difficult to tease out the specific factors that promoted the evolution of social living, since group life carries with it some obvious drawbacks, and the advantages listed will not be the same for all members of the group.

One of the potential disadvantages of acquiring food in a social group is competition from other group members; however, this is likely to be offset by the obvious advantages of foraging in groups. A number of animals can successfully defend a food source, and individuals are able to take advantage of communal knowledge about the location of food. Young group members and smaller individuals who are unable to get at hard nuts and fruits will benefit by being able to pick up bits of food discarded by the others. There appears to be a good correlation between group size and the distribution of food, where species depending on resources that are evenly distributed in small patches tend to live in small groups, while those who depend on large patches of food items, such as figs, that are unevenly distributed, often live in large groups.

There is relatively little hard data on predation in primates, therefore it is difficult to compare the rates of predation on different species and to make assumptions about the relationship between predator pressure and group size. Few primatologists have seen actual instances of predation, since predators often strike at night and tend to avoid the presence of humans. However, the disappearance of animals and the existence of behaviors such as warning calls, displays and vigilant postures are indications that predator pressure is a fact of life in primate society. Although groups of animals are more obvious and visible to predators than single individuals, the advantages of safety in numbers would seem to outweigh this drawback. Clearly, the group has an improved ability to detect predators and can mob and drive off a single predator, whereas a lone animal would be at a greater risk (Fleagle, 1988).

Access to mates does not appear to be a convincing reason for living in groups, since reproduction could be maintained if a male and female met only when the female was in **estrus**. However, given the difference in the reproductive potentials of males and females, the sexes have different strategies to maximize their reproductive success and therefore each benefit differently from group life. The males' ability to impregnate females is limited only by access to estrous females, while the females are limited by time, being able to produce only a finite number of offspring over their reproductive years. Thus the males must compete with other males in order to mate with a number of females while excluding others from mating, and the first priority of females is to choose a fit mate in order to produce healthy infants. It would appear that group living and access to a number of females would be more advantageous for males in terms of reproductive success and of lesser importance for the females (Fleagle, 1988).

In order to ensure that an individual's genetic material is carried over to the next generation, it is crucial that its offspring live to reproductive maturity. Given the female's large, energetic investment in producing an infant, it would be advantageous for her to have assistance in rearing that offspring. Although there is great variety among primate species in the type of assistance available, perhaps the most important factor is kinship and access to close relatives. In large multi-male, multi-female groups, related females live in close proximity to each other and often provide help in rearing an infant, while in monogamous primate societies the care and raising of the infant is shared by the male. Since females in polygynous groups tend to be closely bonded and supportive of each other, it would appear that all types of primate societies have the potential to provide much-needed help for a female raising her offspring. Clearly this would give an added incentive for female primates to live in a social group (Fleagle, 1988).

All of the above **proximate** factors could conceivably contribute to the survival of the individual living in a social group. However, social behavior can also be viewed in terms of evolutionary biology and be said to have evolved through **natural selection.** It is therefore possible that the predisposition of primates to be social and the strong motivation apparent in primates to seek social interactions and form relationships other than kinship are reasons enough to pursue group life (Mendoza, 1984).

Summary

Primatologists have attempted to classify the different types of social organizations found in nonhuman primate species on the basis of ecolgical factors or mating patterns. Classification systems based on habitat use and diet have proven to be ambiguous because of problems encountered in the accurate definition of the variables. The use of mating patterns (monogamy versus polygamy) has provided a more convenient way of describing and categorizing primate social groups. Monogamous family groupings of a male, a female and their offspring is relatively rare in nonhuman primates. Although monogamous species are found in widely separated phylogenetic groups, they have the following characteristics in common: a lack of **sexual dimorphism**, synchrony of activities, territoriality, and little differentiation in the roles of the males and females. Polygamous mating patterns, which are widespread among primates, include polygyny, where one male mates with a number of females, and multi male societies, where a number of males and females mate together. Eisenberg et. al. (1972) have speculated that the different social groups evolved from that commonly found in nocturnal prosimians, in which the range of the male overlaps with the smaller ranges of several females. The theory suggests that the other social systems arose from the increasing tolerance of adult males for females and then for other adult males.

Although group living does have drawbacks in terms of competition for resources from group memebers, the fact that all primates live in some type of a social group indicates that the advantages must outweigh the disadvantages. It has been suggested that the possible factors that have selected for group living are: collective knowledge of sources of food and water, protection from predators by virtue of safety in numbers, greater access to mates, and assistance in raising the young.

Part II

Introduction to the Life-Way Patterns of Nonhuman Primate Species

Chapter 5

Prosimians and Tarsiers

Suborder Prosimii

Traditionally, the division between the prosimians (which included tarsiers) and the anthropoids was made primarily on the basis of sensory systems (Bearder, 1987). Since their appearance on earth, some 50 million years ago, prosimians have occupied a nocturnal niche. And even today they still have the physical adaptations that have allowed them to be highly successful in that niche: large eyes, large mobile ears and a well-developed **olfactory sense**. The tapetum lucidum, a crystalline shield located directly behind the retina, found in all prosimians, reflects light back onto the retina and greatly enhances night vision, while their moist noses are useful in a world where an acute sense of smell is more important than good vision. The fact that the moist nose is fixed to the upper lip, immobilizing it and making their faces relatively expressionless, does not present a great problem when the face cannot be seen.

Although the number of primate field studies increased dramatically in the 1960s, few of these were directed towards prosimians until the late 1970s (Southwick and Smith, 1986). This is not surprising, given their arboreal and nocturnal way of life which presented researchers with two sets of problems. To compound these difficulties, even when prosimian behavior was documented, the information was hard to interpret because the animals live in a completely different sensory world from that of the researchers. Unfortunately, the bias towards the study of species that are more closely related to humans, like Old World monkeys and apes, was very evident in the early days of primate field work (Doyle and Martin, 1974). However, we now realize that there is much to be gained by widening our scope to the investigation of all nonhuman primates, particularly since so many of them are in such jeopardy. Many of the technical difficulties of studying nocturnal species have been overcome, and prosimians are now coming into their own, judging from the number of journal articles devoted to every aspect of their lives.

The prosimians have retained a number of ancestral physical features which have proven to be very effective in allowing them to survive until the present. It must be remembered that although these same features were present in the fossils of their Eocene forebears, it does not mean that prosimians are poorly evolved primates. On the contrary, they are superbly well-adapted to their particular ecological niche, successfully avoiding competition from primates and other animals. The characteristics of modern prosimians that were present in the fossil remains of extinct species are as follows:

1. A dental comb, made up of the incisors and canines of the lower jaw which are joined and project forward to form a scraping mechanism used for grooming and foraging.

2. A grooming claw on the second digit of each foot.

3. A relatively small brain case.

4. Eye sockets that are not completely closed off from the rest of the skull.

5. The lack of a bony tube connecting the inner and outer ear.

Other features that set prosimians apart from the rest of the nonhuman primates are their moist noses, multiple nipples, a **bicornate uterus**, and a less efficient system for nourishing the developing fetus with maternal blood than that seen in tarsiers and anthropoids (Fleagle, 1988). The diminutive size and nocturnal life-style of the modern prosimians living in Africa and Asia has likely been a factor in their successful avoidance of competition from the Old World monkeys and apes.

The following section will provide a general description of the prosimians in the different taxonomic categories and a brief overview of a few representative species. Readers who are interested in the details of the life-way patterns of particular prosimian species are directed to the excellent descriptions in The Complete Guide to Monkeys, Apes and Other Primates (Kavanagh, 1983) and The Natural History of the Primates (Napier and Napier, 1985).

The Suborder Prosimii is divided into two Infraorders: the Lorisiformes and the Lemuriformes.

Infraorder Lorisiformes (See Table 5.1)

The Lorisiformes are made up of one family, the Lorisidae, which has been divided into two distinct subfamilies: the Lorisinae and the Galaginae. These small, nocturnal primates living in the forested regions of Africa and Asia share a number of prosimian features, but differ greatly in their postcranial anatomy and locomotor patterns (Fleagle, 1988).

Subfamily Lorisinae (See Table 5.1) (Four Genera -*Loris*, *Nycticebus*, *Perodictus* and *Arctocebus*)

These prosimians can be recognized by their large eyes, mobile ears, and their distinctive locomotor pattern, which is not seen in any other primate forms. They move slowly through the forest on all four limbs, cautiously climbing to the upper levels of the trees, never leaping, but bridging gaps in the foliage by hand-over-hand progression. This mode of locomotion is ideally suited to capturing the insects that make up at least part of the diet of all lorises. Their hands, with widely divergent thumbs and reduced second digits, provide a powerful gripping mechanism, particularly adapted to their slow, stealthy progression thorough the forest. Since most of the species live in the typical, solitary nocturnal prosimian type of social setting, scent marking is the primary means of communicating information about age, sex, female receptivity and even identity (Napier and Napier, 1985).

Table 5.1 The Infraorder Lorisiformes

Family Lorisidae - Subfamily Lorisinae	
Species	**Common Name**
Perodicticus potto	Potto
Arctocebus calabarensis	Angwantibo, Golden potto
Loris tarigradus	Slender loris
Nycticebus coucang	Common slow loris
Nycticebus pygmaeus	Pygmy slow loris
Family Lorisidae - Subfamily Galaginae	
Galago senegalensis	Senegal bushbaby
Galago gallarum	Somali bushbaby
Galago moholi	South African bushbaby
Galagoides demidovii	Demidoff's bushbaby, Dwarf galago
Galagoides thomasi	Thomas' bushbaby
Galagoides zanzabaricus	Zanzibar bushbaby
Galagoides alleni	Allen's bushbaby
Otolemur crassicaudatus	Large-eared greater bushbaby Fat-tailed bushbaby
Otolemur garnetti	Small-eared bushbaby
Euoticus elegantulus	Needle-clawed bushbaby Western needle-nailed bushbaby
Euoticus inustus	Eastern needle-nailed bushbaby
Euoticus matschiei	Matschie's needle-clawed bushbaby

Sources: Adapted from Napier and Napier, 1985 and Fleagle, 1988.

The Slender Loris (*Loris tarigradus*), living in the forests and woodlands of India and Sri Lanka, is more lightly built and more dependant on a diet of insects than other members of the Lorisinae. The two species of Slow Loris (*Nycticebus*) (see Table 5.1) are widely distributed throughout the the dense jungles of Southeast Asia and Indonesia. These small primates are more robust than the slender loris and have a more diversified

Figure 5.1   Slow loris (*Nycticebus coucang*).
(Courtesy Ruben Kaufman.)

diet of leaves, fruit, birds and birds eggs to supplement the intake of insects. The pottos (Genus *Perodicticus*) are the largest of the lorises and closely resemble the slow loris in their robust appearance. They live in the tropical forests of central Africa, foraging in the upper canopy for the fruit which makes up the major part of there diet. The pottos are well-armed against predators with their bony spines, made up of elongated neck and thoracic vertebrae surrounded by heavy musculature, which present a potential predator with a protective wall of bone and muscle (Napier and Napier, 1985). The impressive power grip of the potto has earned it a reputation for perseverance and endurance which is highly regarded by the African natives.

Subfamily Galaginae *(Four Genera - Galago, Otolemur, Euoticebus and Galigides) (See Figure 5.2)*

Figure 5.2   Senegal galago or Senegal bushbaby
(*Galago senegalensis*). (Photo by Brian Keating.
Courtesy of the Calgary Zoological Society.)

The galagos, or bushbabies, are a diverse group of prosimians found only in forests and woodland savannahs of sub-Saharan Africa. They are small and nocturnal like the lorises, with the large eyes and mobile ears common to nocturnal creatures, but have a very different postcranial anatomy which is adapted to their vertical clinging and leaping style of locomotion. Their elongated hind limbs and tarsal bones, and powerful thigh muscles allow them to make prodigious vertical leaps and propel themselves across wide spaces between trees. Their excellent night vision and hands that that are well-suited to gripping insure a secure landing on the target branch, as well as enabling them to surprise insects and other prey. The different species of galagos exhibit a number of variations of the social system common to most nocturnal prosimians, with the range of the male overlapping the smaller ranges of several females. However, since the males and females tend to go their separate ways, **scent marking** is one of their primary means of communication (Napier and Napier, 1985).

Infraorder Lemuriformes (See Table 5.2)

Members of the Lemuriformes are found only in Madagascar, a large island off the southeast coast of Africa, and occupy a number of very diverse ecological zones, ranging from moist rain forest and seasonally dry forests to the extremely arid country in the southern part of the island. They are the most abundant and diverse of the prosimians, and have the distinction of being the most-studied members of their Suborder. Although most of the 20 to 25 types of lemurs are nocturnal, some of the species have adapted to a diurnal lifestyle. Since it has been established that Madagascar separated from mainland Africa during the reign of the dinasaurs some tens of millions of years before the appearence of primate-like mammals, the lemurs must have rafted across the Mosambique Channel on islands of vegetation around 50 milllion years ago. The ancestors of the modern lemurs were presented with a wealth of ecological potentials and, in the absence of predators and competitors, underwent an adaptive radiation resulting in many more species than we see today (Tattersall, 1993). It is estimated that in the last 2,000 years, since the arrival of humans on the island, 14 species have become extinct, and at the present time, all of the extant species are on the endangered list, primarily due to widespread habitat destruction (Fleagle, 1988).

The lemurs are very diversified, not only in the ecological niches the inhabit, but in size and in other physical attributes. The tiny mouse lemur weighs about 60 gms (2 ounces), while the indri weighs up to 10 kg (22 pounds). The Infraorder Lemuriformes is divided into two Superfamilies: the Lemuroidea and the Daubentonioidea. There are four families of Lemuroidea: The Cheirogalidae, with four genera; the Lemuridae, which includes the largest number of genera and species; the Lepilemuridae; and the Indriidae, composed of three genera. The superfamily Daubentonioidea includes one family, the Daubentoniidae, which contains only a single genera and species, the highly endangered aye-aye (Richard, 1987; Fleagle, 1988).

Family Cheirogalidae *(Four genera - Microcebus, Cheirogaleus, Allocebus and Phaner)*

Although they are among the smallest primates, the three species of Mouse Lemur

Table 5.2 The Infraorder Lemuriformes

Family Cheiogaleidae	
Species	**Common Name**
Microcebus murinus	Grey mouse lemur, Grey lesser mouse lemur
Microcebus rufus	Brown mouse lemur, Brown lesser mouse lemur
Microcebus coquereli	Coquereli's dwarf lemur, Coquereli's mouse lemur
Cheirogaleus major	Greater dwarf lemur
Cheirogaleus medius	Fat-tailed dwarf lemur
Phaner furcifer	Fork marked dwarf lemur
Allocebus trichotis	Hairy-eared dwarf lemur
Family Lemuridae	
Lemur catta	Ring-tailed lemur
*Lemur fulvus**	Brown lemur
Lemur mongoz	Mongoose lemur
Lemur macaco	Black lemur
Lemur rubriventer	Red-bellied lemur
Lemur coronatus	Crowned lemur
Varecia variegata	Ruffed lemur
Hapalemur griseus	Gentle bamboo lemur
Hapalemur simus	Greater bamboo lemur
Hapalemur aureus	Golden bamboo lemur
Family Lepilemuridae	
Lepilemur mustelinus	Sportive lemur
Family Indriidae	
Indri indri	Indri
Propithecus verreauxi	White sifaka, Verreauxi's sifaka
Propithecus diadema	Diademed sifaka
Avahi laniger	Woolly lemur
Family Daubentoniidae	
Daubentonia madagascariensis	Aye-aye

*Subspecies names not included
Sources: Adapted from Napier and Napier, 1985 & Fleagle, 1988.

(genus *Microcebus*) are highly successful nocturnal predators. They will eat almost anything, and scurry around their tropical forest habitat searching for fruit, flowers, insects, small birds and reptiles. They have the large eyes and mobile ears of other nocturnal prosimians, and, although they tend to forage alone, the females are relatively gregarious, sharing their sleeping nests with other females and their young.

The fat-tailed dwarf lemur (*Cheirogaleus medius*), one of the three species of dwarf lemur, lives in the dry forest of southern Madagascar and has the distinction of being the only primate that is a true hibernator. Since the fruit, flowers and nectar that make up their diet are scarce during the dry season, these small nocturnal creatures hibernate for six to eight months, living off reserves of fat stored in their tails. Very little is known about the social pattern of the dwarf lemurs, but they are thought to live in the social system common to other nocturnal prosimians (Napier and Napier, 1985).

Family Lemuridae *(Three Genera - Lemur, Varecia and Hapalemur)*

Members of the Genera *Lemur,* the best known group in this subfamily, are the most widespread and diverse in habitat and social pattern of any of the prosimians, and are often referred to as "true lemurs." They are all diurnal, with at least some species living in all of the forested areas of Madagascar (Napier and Napier, 1985), and are more like monkeys in their social systems and ecological adaptations than any of the other prosimians. Unlike the nocturnal prosimians, their diet consists mainly of fruit, seeds, flowers, leaves and very few insects. It appears that, although they retain some of the physical adaptations to

Figure 5.3 Ring-tailed lemur (*Lemur catta*). Note the luxuriant striped tail. (Courtesy Ruben Kaufman.)

Figure 5.4 A group of ring-tailed lemurs in a typical sunning position. (Courtesy Ruben Kaufman.)

nocturnal living, these prosimians were able to take up a diurnal life-style in the absence of competition from monkeys and apes (Kavanagh, 1983). Many species of lemurs have short arms and elongated hind-limbs, and are physically adapted for supplementing their general quadrupedal locomotor style with clinging and leaping (Fleagle, 1988).

One of the most distinctive looking and certainly the most intensively studied of the prosimians are the ring-tailed lemurs (*Lemur catta*) (see Figures 5.3 and 5.4). They are lightly-built animals about the size of a domestic cat, with long hind legs, short arms and the magnificent, long, plumed black and white tails that give them their name. Their tails are often carried aloft and communicate both visual and olfactory signals. They live in large, cohesive multi-male, multi-female groups in the dry southern forests of Madagascar. They are predominantly semi-terrestrial, traveling quadrupedally along the ground, but can also be found foraging and feeding in all levels of the forest, and are extremely adept at vertical clinging and leaping through the trees (Richard, 1985). Although they are active in the daylight hours, their vision is not highly developed and their moist noses and scent glands attest to the importance of olfaction in their communication system (Tattersall, 1982).

L. catta are extremely gregarious animals, whose social system rivals that of some of the Old World monkeys in complexity. The social group centers around the adult females, who spend much of their leisure time grooming each other or caretaking the infants of other females. Although well-defined dominance hierarchies have been recognized in males and females, the females are generally dominant over the males, particularly in regard to access to preferred foods. Studies indicate that the females spend their entire life in their natal group, while the males show a widespread tendency to transfer between groups (Budnitz and Dainis, 1975). Most of the competition between males and contact between males and females take place during the mating season, which lasts for two weeks every year.

The other five species of lemurs have many of the same features in physical attributes and locomotor pattern. However, while the male and female *L.Catta* look very much the

Figure 5.5 Female black lemur (*Lemur macaco*). The males of the species are totally black. (Courtesy Ruben Kaufman.)

Figure 5.6 Female brown lemur (*Lemur fulvus rufus*) with a male infant. The sexual dichromatism is evident in head color, the male having a red cap. (Courtesy Ruben Kaufman.)

same, the sexes in the other species show distinct differences in coloring and adornment. The most striking example are the black lemurs, (*Lemur macaco)* where the males are completely black and the females are brown with white ear tufts (see Figure 5.5).

There are a number of subspecies of brown lemurs (*Lemur fulvus*) (see Figures 5.6 and 5.7) who are widely distributed in the forests throughout Madagascar. The diferent types are sexually dichromatic, and the males and females display a variety of colorful coat patterns. Although the brown lemurs are similar in size and locomotor pattern to the ring-tailed species, they are totally arboreal and subsist primarily on a diet of leaves. Although their multi-male, multi-female social structure is found in many *Lemur* species, the usual pattern of female dominance is not apparent (Pereira et al., 1990). Although *Lemus fulvus* and *Lemur mongoz* (see Figure 5.8) are primarily diurnal creatures, there is

Figure 5.7 Male brown lemur in the leaf litter (*Lemur fulvus rufus*). (Courtesy Ruben Kaufman.)

Figure 5.9 Crowned lemur (*Lemur coronatus*). (Courtesy Ruben Kaufman.)

Figure 5.8 Female mongoose lemur (*Lemur mongoz*). (Courtesy Ruben Kaufman.)

some indication that are are active at night under some conditions. The ecology and behavior of the other species in the Lemur genus is less well known. However, with the increase in the numbers of primatologists working in Madagascar, this situation is bound to improve (see Figure 5.9).

The ruffed lemur, *Varecia variagata* (see Figure 5.10), the only species in the genus *Varecia*, live in large, widely-dispersed groups that tend to break up into small foraging parties. It is interesting to note that they are the most frugivorous of the Malagasay lemurs and exhibit the same fission-fusion type of social system common to spider monkeys and chimpanzees, who are also fruit-eating primates. They are particularly noted for their vocal repertoire of loud, raucous roars and barks, which appear to serve in intergroup communications, as well as predator alarms and mating displays (Periera et al., 1988).

There are three species of Hapalemurs, the third genera of the Lemuridae family. The small gentle lemur (*H. griseus*) (see Figure 5.11) has a greater distribution than the rarer greater bamboo (*H. simus*) and golden bamboo lemurs (*H. aureus*). As the name would suggest, these primates live in the bamboo forests, and subsist entirely on various parts of the the bamboo plant. The diminishing habitat of these cryptic animals, who seem to be neither totally diurnal or nocturnal, has placed them in extreme danger of extinction, at least in the wild. Perhaps the best chance for the survival of the gentle lemurs, the aye-aye and a number of other lemur species living in a small area of the rainforests in southeast Madagascar, has been the establishment of the Ranomafana National Park in 1991. The success of this park in terms of the protection of the habitat and serving the needs of the people living in the adjacent areas will be critical to the formation of other nature reserves in Madagascar (Wright, 1992).

Figure 5.10   Ruffed lemur (*Varecia variegata*). (Courtesy Ruben Kaufman.)

Figure 5.11   Grey gentle lemur (*Hapalemur griseus*). (Courtesy Ruben Kaufman.)

Family Lepilemuridae *(Genus Lepilemur)*

Lepilemurs, or sportive lemurs, spend their days sleeping in the cool hollows of trees, coming out at night to forage for the leaves that make up the bulk of their diet. They are physically and behaviorally well-adapted for survival on their fibrous, nutrient-poor diet. The leaves are partially digested in their capacious lower bowel, after which the feces are reingested in order to extract every bit of good from the remains. They are vertical clingers and leapers, whose small ranges and energy-conserving activity patterns, allow them to be very successful in many of the forested areas of Madagascar (Napier and Napier, 1985).

Family Indriidae *(Three genera - Propithecus, Indri and Avahi)*

The members of this family of prosimians are exclusively vertical clingers and leapers, with the typical long, strong hind limbs, short arms, erect trunks, and prehensile hands and feet. Of the three genera, only the *Avahi* is nocturnal.

The sifaka (Genus *Propithecus*) (see Figure 5.12), named for their distinctive vocalization, are very adaptable animals who are able to live in the eastern rain forests as well as the dry, deciduous forests of western Madagascar. Although their vegetarian diet consists of kily pods, seeds, bark and flowers which are available in both types of forest, their ability to exploit a dry habitat depends on their apparent skill in extracting their fluid needs from cacti and other succulents. Sifakas spend most of their time leaping gracefully from tree to tree. However, their short arms and long legs, which serve them well in the

Figure 5.12 A group of Verreauxi's sifakas (*Propithecus verreauxi*). (Courtesy Ruben Kaufman.)

Figure 5.13   The aye-aye (*Daubentonia madagascariensis*). (Courtesy David Haring.)

trees, make movement on the ground difficult, and they have to resort to an ungainly, bipedal hop. They are sociable animals who live peacefully in small, stable multi-male, multi-female groups. Only the annual mating season, when the males compete fiercely for the females, disrupts the harmony of the group.

The Indri (*Indri indri*), the largest of the prosimians, spend all of their waking hours, high in the dense rain forests of the northeast coast of Madagascar. Although they are relatively unobtrusive to the eye, their whereabouts can be detected by their loud, barking choruses. Their diet is made up primarily of leaves, and their activity pattern is similar to that of other folivores: long periods of foraging and resting broken by the occasional territorial displays. Unlike other prosimians, the indris live in monogamous family groups who advertise their small territories with vocal displays. The female appears to be totally dominant over the male. She and her offspring feed from the prime areas high in the canopy, whereas the male has to make do with the available lower leaves (Napier and Napier, 1985).

Family Daubentoniidae

The only member of the Family Daubentoniidae and the genus *Daubentonia*, the aye-aye (*Daubentonia madagascariensis*), is the most unusual of all the prosimians, both in appearance and physical features (see Figure 5.13). Its ghostly countenance and nocturnal life-style have given rise to superstitions about its evil nature, leading the people of Madagascar to fear and persecute them. The aye-aye is large for a nocturnal primate, and have a fox-like appearence with a pointed, moist nose, large ears and a bushy coat and tail. Their unique features, the 18 sharp teeth which grow continuously, and the long, sinister, wire-like middle fingers are admirably suited to securing them their diet of fruit and small insects. Their sharp teeth can scrape away bark to get at beetle grubs, while their middle fingers are ideal for puncturing eggs, probing for insects and digging deep into coconuts. It is often suggested that, in its quest for wood-beetle grubs, the aye-aye seems to have taken the place of the woodpecker, which is not found in Madagascar. More recently, Iwano and Iwakawa (1988) have argued that the aye-aye's ability to extract the meat from nuts allows them to fill the niche occupied by squirrels in other areas of the world. Their social lives are somewhat of a mystery, as the few remaining animals in the wild have not been well-studied. Although the total primate population on Madagascar is endangered to some degree, the aye-aye is the closest to complete extinction, with an unknown number of animals alive in the wild. Perhaps the only salvation for these unique primates is the possibility of breeding them in captivity.

Suborder Tarsioidea (See Table 5.3)

The primates in the genus *Tarsius*, the tarsiers, are the only representatives of the Suborder Tarsioidea (see Figure 5.14). They are tiny, nocturnal creatures that were once widespread throughout North America, Europe and Asia, but now can only be found in the tropical forests of southeast Asia (Fleagle, 1988). Traditionally, they were included in the Suborder Prosimii because of their many prosimian-like features; however, it seems

Table 5.3

Suborder Tarsioidea - Family Tarsiidae	
Species	**Common Name**
Tarsius syrichta	Phillipine tarsier
Tarsius bacanus	Bornian tarsier
Tarsius spectrum	Spectral tarsier
Tarsius pumilus	Pygmy tarsier

Source: Adapted from Fleagle, 1988

Figure 5.14 The Bornian tarsier (*Tarsius bacanus*). Note the enormous eyes relative to the size of the body. (Courtesy David Haring.)

that the more we know more about them, the more confusing their proper taxonomic position becomes. Although modern tarsiers inhabit a nocturnal niche, their eyes are adapted to daytime vision, with the presence of a **retinal fovea** (the light sensitive part of the retina), and the absence of the **tapetum lucidum** behind the retina that acts to reflect light back onto the retina. Tarsiers retain some of the primitive features found in prosimians: an unfused lower jaw, grooming claws on the second and third toes, multiple nipples, mobile ears and a bicornate uterus. However, their many anthropoidal characteristics make it difficult to classify them with the lorises and lemurs. Perhaps the most striking feature that separates the tarsier from prosimians is the dry nose similar to that found on monkeys and apes. Other traits common to tarsiers and anthropoids are:

1. The major blood supply to the brain is through the promontory branch of the **carotid artery.**

2. A bony tube that links the inner and outer ear

3. The attachment of the fetus to the uterine wall, where the blood vessels make direct contact with those of the mother, allowing for an efficient transference of nutrients from maternal to fetal blood.

4. Complete bony eye sockets.

5. Lack of a dental comb.

Confusing the taxonomic picture even more, the tarsiers have some unique features found in no other primate species. They have only 34 teeth, compared to the 36 teeth found in all other prosimians except the aye-aye, with only one incisor in the lower jaw, and a fused tibea and fibula providing them with added ability to propel themselves through the air (Fleagle, 1988). This combination of prosimian and anthropoidal traits is compatible with two equally plausible views of primate evolution. One suggestion is that the early tarsiers were prosimians that eventually evolved into anthropoidal forms. A more recent argument is that the tarsiers and anthropoids evolved from a common diurnal ancestor, after which the tarsier reverted to a nocturnal lifestyle. In any event, one solution to the taxonomic dilemma, which admittedly is not universally accepted, is to place the tarsiers in their own suborder, serving to recognize the similarities to the other suborders, as well as their own unique attributes.

Four species of tarsiers have now been recognized in different areas of southeast Asia:

Tarsius bacanus (Borneo tarsier), found in Borneo, Java and Sumatra;
Tarsius spectrum (Spectral tarsier), from Sulawesi;
Tarsius syrichta (Phillipine tarsier), from the Phillipines; and
Tarsius pumilus (formerly a subspecies of the spectral tarsier), found in the mountain forests of Sulawesi (Fleagle, 1988).

Tarsiers are unusual looking creatures with enormous eyes, and inordinately long hands and hind limbs in proportion to their tiny bodies and long, hairless tails. They are among the most efficient vertical clingers and leapers of the primate world, with their enlarged tarsal bones, long hind limbs and grasping hands that enable them to cover distances of 6m (20 ft) with a single leap. Although their eyes do not move in their sockets, their heads can swivel 180 degrees, allowing them to surprise their prey much in the manner of owls. Unlike other prosimian species, tarsiers are voracious foragers who subsist totally on animal matter. They use their prodigious leaping ability to capture snakes, insects, lizards and small birds, killing them with a single bite and eating them completely: beaks, feathers and all (Napier and Napier, 1985).

The social life of tarsiers is not well known, but it is suspected that there is considerable interspecific variation in their social systems (Fleagle, 1988). There is evidence that spectral tarsiers live in monogamous family groups that defend their small territories vigorously against other conspecifics (MacKinnon and MacKinnon, 1980). Other species, such as *T. bacanus*, live in the social system common to other nocturnal prosimians, where the range of the solitary male overlaps with those of a number of females (Compton and Andau, 1987). All species communicate by means of scent-marking and a variety of high-pitched vocalizations.

The gestation period of all tarsiers is 180 days, similar to that of Old World monkeys and the well-developed infants are large in comparison to the mother, weighing about 30% of her weight at birth. Surprisingly, there is no evidence of male or female alloparenting to ease the burden of raising such large infants. The relatively advanced state of physical development of the new born tarsiers, however, speeds their maturation, and by four weeks of age, they are able to hunt for their own food (Napier and Napier, 1985).

Chapter 6

The New World Monkeys

Suborder - Anthropoidea Infraorder - Platyrrhini

New World monkeys are found in Central and South America, distributed from central Mexico to southern Argentina where their range is cut off by the Andes mountains and the pampas zone. They can be found wherever there are tropical forests and are often referred to as Neo-tropical monkeys. Since there is no fossil evidence of primates in South America before the Oligocene Epoch, the New World monkeys have been evolving on their own for over 30 million years.

As was the case with prosimians, New World monkeys seemed to be low on the priority list for field studies, a situation reflected in the relatively few early publications coming from South America, compared to those from Africa and Asia. This was partially due to their phylogenetic distance from us, as well as their arboreal lifestyle which makes study more difficult (Southwick and Smith, 1986). This situation has improved dramatically, and since 1980, interest in New World species has grown as primatologists have realized the importance of expanding our knowledge to arboreal forms in order to fully understand primate life-way patterns. This is especially critical now due to the widespread habitat destruction in the South American countries where the Neo-tropical monkeys exist and the threat of extinction is imminent.

Although New World monkeys look very much like Old World monkeys and live in social systems that are equally complex, there are a number of physical features that differentiate the two primate forms (see Table 6.1). The infraorder designation, Platyrrhini, is derived from aspects of their nasal anatomy and refers to the flat noses and the sideways orientation of the nostrils. Old World monkeys and apes belong to the infraorder Catarrhini, a term that describes their downward directed nostrils. New World species are medium-sized animals with relatively conservative limb proportions well-suited to their arboreal niche. Although all species have tails, in some genera the tail is prehensile, with a finger tip-like pad, and acts as a fifth limb. Members of the Platyrrhini exhibit primitive dental and cranial features not found in the Catarrhini, which include the retention of a third premolar and the lack of a bony tube linking the inner and outer ear (Fleagle, 1988).

Primatologists have difficulty agreeing on the taxonomy of New World monkeys below the level of the superfamily (Ceboidea), resulting in a confusing number of different classifications at the family and subfamily level. The most convenient system, although not necessarily the most accurate, is to divide the Ceboidea into two families: the Callitrichidae and the Cebidae.

Table 6.1 Features distinguishing New World from Old World Monkeys

New World Monkeys	Old World Monkeys
1. Flat noses, with sideways-directed nostrils	1. Downward-directed nostrils

2. $\underline{2.1.3.3}$ Dental formula
  2.1.3.3   3 premolars,
       36 teeth in all
	2. $\underline{2.1.2.3}$ Dental formula
	  2.1.2.3 2 premolars,
	       32 teeth in all

3. Some species with prehensile tails with fingertip-like pads	3. No prehensile tails
4. No ischial callosities	4. Ischial callosities
5. Arboreal life-style	5. Semi-terrestrial and arboreal life-style
6. Relatively small-bodied	6. Large and small-bodied
7. Ear drum attached to the tympanic ring	7. Ear drum attached to a bony tube linked to the outer ear

Family Callitrichidae (Four genera - *Callithrix, Cebuella, Saguinus* and *Leontopithecus*) (See Table 6.2)

These colorful, exotic looking primates are the smallest of all monkey species and are unusual in that they have claws on all digits except the big toe. The two major groups in this family are the marmosets and the tamarins which are found deep in the forests of Central and South America. They are very difficult to study in the wild, due their diminutive size, their retiring nature and the fact that they are exclusively arboreal and rarely come to the ground. However, they breed well in captivity and much of what we know about them comes from laboratory and zoo studies (see Figures 6.1 and 6.2).

All members of the Callitrichidae are omnivorous, eating fruit, leaves and insects. However, tree gum forms an important part of the marmoset diet. This fondness for gum could explain the "short-tusked" nature of marmosets' canine teeth, which enhances their ability to scrape through tree bark. In the tamarins, as in most other nonhuman primate species, the canines extend well beyond the incisors (Napier and Napier, 1985:107). They are quadrupedal branch runners, are able to use their claws to climb vertical branches, and resemble squirrels in their rapid and somewhat jerky movements through the trees (Napier and Napier, 1985).

Most species of marmosets (Genera *Callithrix* and *Cebuella*) and tamarins (Genus *Saguinus* and *Leontopithecus*) live in monogamous family groups consisting of an adult male,

Table 6.2 The Infraorder Platyrrhini - Family Callitrichidae

Species	Common Name
Cebuella pygmaea	Pygmy marmoset
Callimico goeldii	Goeldi's marmoset, Goeldi's monkey
Callithrix argentata	Silvery marmoset, Blue-tailed marmoset
Callithrix humeralifer	Tassel-eared marmoset
Callithrix jacchus	Common marmoset
Callithrix aurita	Buffy tufted-ear marmoset
Callithrix flaviceps	Buffy-headed marmoset
Callithrix geoffroyi	White-faced marmoset
Callithrix penicillata	Black tufted-ear marmoset
Saguinus nigricollis	Black and red tamarin
Saguinus fuscicollis	Saddle-back tamarin
Saguinus mystax	Moustached tamarin
Saguinus labiatus	White-lipped tamarin
Saguinus imperator	Emperor tamarin
Saguinus midas	Red-handed tamarin, Negro tamarin
Saguinus inustus	Inustus tamarin
Saguinus bicolor	Barefaced tamarin, Pied tamarin
Saguinus oedipus	Cottontop tamarin
Saguinus leucopus	White-footed tamarin
Leontopithecus rosalia	Golden lion tamarin
Leontopithecus chrysemolas	Golden-headed lion tamarin
Leontopithecus chrysopygus	Golden-rumped lion tamarin

Sources: Adapted from Napier and Napier, 1985 and Fleagle, 1988.

an adult female and a number of offspring. Since the females give birth to twins twice yearly, the social groups tend to expand rapidly. In many respects, these primates represent textbook examples of monogamous societies. They show little sexual dimorphism in size or coloring, the activities of the adults are synchronized, their territories are actively defended by both males and females and the males take a very active role in caring for the infants. It has been reported that as soon as the infants are born, the males take over many of the caretaking duties, returning them to the mother only for nursing. It is also common for the older offspring to take a prominent part in raising the young callitrichids, carrying them, playing with them and even sharing their food. The sexual maturity of the offspring appears to be suppressed while they remain with the family group, and only the adult female gives birth to the twins twice yearly. At some point, the adult female becomes fiercely agonistic towards her oldest daughters and the male becomes intolerant towards

Figure 6.1   Common marmoset (*Callithrix jacchus*). (Photo by Brian Keating. Courtesy of the Calgary Zoological Society.)

Figure 6.2   Cottontop tamarin (*Saguinus oedipus*). (Photo by Brian Keating. Courtesy of the Calgary Zoological Society.)

his oldest sons, at which point they leave the family and begin their own social group (Goldizen, 1987). The males and females of both the marmosets and tamarins show few affiliative gestures towards each other and, since there is little evidence of pair-bonding, their approach towards raising a family has been likened to that of a business venture.

True marmosets are found from the Amazon Basin to Eastern Bolivia, while tamarins are widely distributed throughout the tropical forests of Central and South America as far as the Amazon Basin. The only marmoset that is sympatric with the tamarins is the pygmy marmoset (*Cebuella pygmaea*). Despite the distinction of being the smallest monkey in the world, the pygmy marmoset is successful in the wild because it is able to exist in disturbed areas as long as there are small patches of forest with enough trees to satisfy its staple diet of gum and sap. The presence of the pygmy marmoset can be easily detected by numerous, characteristic puncture marks on the tree trunks (Terborgh, 1983).

Goeldi's marmosets (*Callimico goeldii*), sometimes referred to as Goeldi's monkeys, are classified with the Calltrichids even though they have dental features and the single births characteristic of the other Platyrrhines. In all other respects, such as the claws on some digits, limb proportions and lack of sexual dimorphism, they resemble the other marmosets and tamarins. *Callimico* live in low bushes and bamboo patches in the upper regions of the Amazon, where they forage for fruit and a large variety of invertebrates. As with the other Callitrichids, the male in the monogamous family group takes an important role in the care and raising of the young (Napier and Napier, 1985; Fleagle, 1988).

There is evidence that in at least one species, the saddle-back tamarin (*Saguinus fuscicollis*), there is some deviation from the monogamous social system common to most Callitrichidae. Terborgh and Goldizen (1985) report that some groups of saddle-back tamarins live in polyandrous groups in which the adult female mates with more than one male. The beautiful golden lion tamarin (*Leontopithecus rosalia*) is one of the most endangered species of New World monkeys, partly because of its desirability as a pet, but

Figure 6.3   Golden lion tamarin (*Leontopithecus rosalia*). (Photo by Brian Keating. Courtesy of the Calgary Zoological Society.)

primarily because of the destruction of its habitat (Napier and Napier, 1985) (see Figure 6.3). An effort is being made to breed these charming animals in captivity and to introduce them carefully into some of the few remaining Brazilian forests where they can be found in the wild (Mittermeier and Cheney, 1987).

Family Cebidae *(Genera Callicebus, Cebus, Ateles, Alouatta, Saimiri, Pithecia, Chiropotes, Cacajao, Aotus, Lagothrix, and Brachyteles)* (Table 6.3)

The Family Cebidae is a large and diversified group of New World monkeys that has been divided into anywhere from five to seven subfamilies that vary in composition, depending on the source. In order to avoid this taxonomic confusion, the following discussion will be confined to several species representative of the Cebidae family, without reference to their subfamily grouping.

Titi Monkeys *(Genus Callicebus)*

The titi monkeys (Genus *Callicebus*) are small monkeys with short faces, furry bodies and long tails that live deep in the mangrove swamps of the South American river basins. Their somber coloring, shy, retiring nature and habit of freezing when confronted with danger, in combination with their inhospitable habitat, makes them exceedingly difficult to study in the wild. As with the Callitrichids, much that is known about them comes from laboratory or zoo studies.

They live in monogamous family groups which are much smaller than those of the Callitrichids, since the females only bear one infant yearly, and unlike the marmosets and

Table 6.3 Infraorder Platyrrhini - Family Cebidae

Species	Common Name
Callicebus moloch	Dusky titi monkey
Callicebus personatus	Masked titi monkey
Callicebus torquatus	Yellow-handed titi monkey
Aotus trivergatus	Douroucouli, Owl monkey, Night monkey
Pithecia pithecia	White-faced saki
Pithecia irrorator	Bald-faced saki
Pithecia monachus	Monk saki
Pithecia aequatorialis	Equatorial saki
Pithecia albicans	White saki
Chiropotes satanas	Black-bearded saki
Cacajao calvus	Bald uakari, White uakari
Cacajao melanocephalus	Black-headed uakari
Cacajao rubicundus	Red uakari
Cebus apella	Tufted capuchin, Black-capped capuchin
Cebus albifrons	White-fronted capuchin
Cebus capucinus	White-throated capuchin
Cebus nigrivitattus	Wedge-capped capuchin, Weeper capuchin
Saimiri sciureus	Common squirrel monkey
Saimiri oestedii	Red-backed squirrel monkey
Alouatta belzbul	Black and red howler
Alouatta fusca	Brown howler
Alouatta palliata	Mantled howler
Alouatta seniculus	Red howler
Alouatta villosa	Guatemalan howler
Alouatta caraya	Black howler
Ateles belzebuth	Long-haired spider monkey
Ateles fusciceps	Brown-headed spider monkey
Ateles geoffroyi	Black-handed spider monkey
Ateles paniscus	Black spider monkey
Brachyteles arachnoides	Woolly spider monkey
Lagothrix flavicauda	Yellow-tailed woolly monkey, Hendee's woolly monkey
Lagothrix lagothrica	Common woolly monkey, Humboldt's woolly monkey

Sources: Adapted from Napier and Napier, 1985 and Fleagle, 1988

Figure 6.4   The owl monkey or douracouli (*Aotus trivirgatus*). The only nocturnal species of monkey. (Photo by Brian Keating. Courtesy of the Calgary Zoological Society.)

tamarins, the mated pair are incredibly bonded. They are almost always in contact, grooming each other, holding hands and sitting close together with their tails entwined. The strength of the pair-bond is particularly apparent in the degree of stress shown when the pair are separated. The titi monkeys exhibit many of the same characteristics typical of other monogamous primate types. The males and females are identical in appearance, and go about their daily activities in close proximity, foraging and feeding on fruit and leaves and vocalizing loudly to advertise and defend their small territories together. The male in the family group takes a great deal of interest in the care and raising of the infants, carrying them around on his back, playing with them and even allowing them to beg for food (Mason, 1975).

Owl Monkeys *(Genus Aotus)*

Another member of the Cebidae that lives in monogamous family groups is the owl monkey (*Aotus trivigatus*), the only nocturnal member of the Anthropoidea (see Figure 6.4). The douricouli, as it is commonly called, likely evolved from a diurnal ancestor, since it lacks the adaptations to night vision found in the nocturnal prosimians. They are primarily fruit-eaters and are able to live successfully in many different types of forests from Central America to Argentina. As with other monogamous species, the owl monkeys appear to be very territorial, defining their ranges by scent marking and defending them with a repertoire of vocalizations (Napier and Napier, 1985).

Figure 6.5   The white uakari (*Cacajao calvus calvus*). (Photo by Russell A. Mittermeier, Conservation International.)

Sakis and Uakaris (Genera *Pithecia and Cacajao*) *(Figure 6.5)*

The sakis (genus *Pithecia*) are the most distinctive of the New World monkeys in terms of their physical appearance. They are widely distributed throughout the Amazon Basin in the tropical forest areas that are not prone to flooding. This allows them to move easily in the lower canopy and the understory of the forests. Their unusual dental pattern, which is characterized by large procumbent incisors and small premolars and molars, is possibly an adaptation to their diet of seeds and fruit (Fleagle, 1988). Although they are often found in small monogamous social groups, Fleagle (1988) has suggested that the incidence of larger aggregations may be indicative of a fission-fusion type of social organization.

Although little is known about the behavior and social patterns of the uakari (genus *Cacajao*), which is closely related to the Saki, it is worth mentioning because of its grotesque appearance. Of the three species, the red uakari is the most familiar, with its red face, hairless scalp, shaggy red coat and bushy tail. Uakaris seem to prefer a wet habitat, as they live in large groups in the seasonally flooded forests of the Amazon Basin. The few field studies indicate that they move quadrupedally through the trees and use all four limbs in suspensory postures while foraging for there staple diet of fruit (Fontaine, 1981).

Capuchins *(Genus Cebus)*

Perhaps the best known group of the Cebidae family are the capuchins (Genus *Cebus*). There are four recognized species that are widely distributed in Central and South America as far south as Argentina (see Figure 6.6). They are medium-sized animals with long, semi-prehensile tails that live in somewhat dispersed multi-male, multi-female groups in the middle and lower canopy of the forest. They are quadrupedal branch runners who spend much of their day foraging and feeding for the fruit and insects that make up the major part of their diet (Terborgh, 1983). Kavanagh (1983:98) refers to them as "the hustlers of the American primate world," a title they richly deserve, as they are deliberate, manipulative and destructive foragers using many inventive strategies to extract embedded and difficult-to-get food items (Izawa, 1979; Fedigan, 1990). It has been suggested that, among nonhuman primates, such complex foraging strategies are only found in cebus monkeys and chimpanzees (Parker and Gibson, 1977). Cebus monkeys are popular subjects for laboratory learning research, and it is suggested that their foraging skill contributes to their proficiency in tests requiring manipulative ability.

All cebus species appear to have both male and female dominance hierarchies. However, it is often the dominant male who leads the group movements, acts to protect the group and is the center of group activities. The adult males tend to transfer between groups, while the females remain in their natal groups for life (Izawa, 1980).

Squirrel Monkeys *(Genus Saimiri)*

Squirrel monkeys (Genus *Saimiri*) are among the smallest members of the Cebidae, but are less like squirrels in size and locomotor patterns than the marmosets and tamarins

Figure 6.6   Two subadult white-throated capuchins (*Cebus capucinus*) in Santa Rosa National Park, Costa Rica. (Courtesy Katherine C. MacKinnon.)

(see Figure 6.7). They live in large groups in all of the major forest types in Central and South America. They are arboreal animals, that move quadrupedally, primarily in the understory of the forest with an occasional trip to the forest floor to forage (Fleagle et al., 1981). Foraging and feeding take up the major portion of their daily activities, as they are constantly on the move searching for insects in the foliage. They are omnivorous, eating fruit and other vegetable matter, but they prefer insects if available. Since they are unable to crack hard nuts and tough seed pods, they rely on their speed and agility to make up for their lack of strength (Fleagle et al., 1981). Squirrel monkeys can often be found traveling in polyspecific groups with cebus monkeys, scavenging food and taking advantage of the wasteful eating habits of the larger and stronger monkeys (Terborgh, 1983).

All group members except adult males are attracted to the adult females, who are the focal point for group activities. Except at mating season, when the adult males seek out the females, all of the social activity occurs within the same age/sex classes. During the mating season, the males build up subcutaneous fat deposits and actively compete for estrous females. They are among the least social of primate species, grooming each other rarely, but although they may avoid social contact, group members appear to monitor each other constantly. Squirrel monkeys, like cebus monkeys, have been used extensively in laboratory learning tests and are able to master most of the tasks set for them. However, they are hampered in those requiring time and concentration by their short attention span and lack of motivation to complete the task (Fragaszy, 1985).

Howler Monkeys (Genus Alouatta)

Howler monkeys (Genus *Alouatta*) are found in tropical forests from the northern to the southern limits of primate habitation in Central and South America and are the most widely distributed of the New World monkeys (see Figure 6.8). Their success is due to their

Figure 6.7   A common squirrel monkey (*Saimiri sciureus*). (Photo by Brian Keating. Courtesy of the Calgary Zoological Society.)

Figure 6.8   A mantled howler monkey (*Alouatta palliata*) in Santa Rosa National Park, Costa Rica. The monkey had been tranquilized in order to be tagged. (Courtesy Francois LaRose.)

ability to survive in refuge forests with only a few clumps of trees as well as the fact that they are unpalatable and left alone by hunters. Howlers are large, robust animals who live primarily on leaves. Although they do not have any physical adaptations for the digestion of a fibrous, nutrient-poor diet, their activity patterns are well-suited for survival as a folivore. They tend to eat quantities of young leaves rich in protein and sugars, and then spend long periods resting and digesting their food. Their mode of slow, ponderous quadrupedal travel through the trees is geared to save as much energy as possible, and their prehensile tail, with its fingerprint-like pad, acts as an efficient fifth limb allowing them to span gaps in the foliage without leaping (Crockett and Eisenberg, 1985).

The males are about 20% larger than the females and can be easily distinguished by the prominent bulge below the chin which houses the large hyoid bone. This forms a hollow resonating chamber, lending added volume to their booming vocalizations. The resounding roar that gives them their name advertises the location of their groups, and functions as a spacing mechanism. However, their choruses can be heard at any time of day, seemingly for a variety of reasons. Although *Alouatta palliata*, the species that has been studied the most, often lives in multi-male, multi-female groups with one breeding male and his male offspring, there appears to be some intraspecific, as well as interspecific variability in group size and composition. The adult females frequently emigrate from their natal groups; consequently, female howlers are very competitive and do not show the same degree of cohesion found in groups with male dispersal patterns (Crockett and Eisenberg, 1985).

Spider Monkeys (Genus *Ateles*)

Spider monkeys (Genus *Ateles*) are distributed from southern Mexico to the Amazon Basin. However, because of their need for large stands of primary forest to accommodate their extensive home ranges, and the fact that they have been continually hunted for food, a number of species are on the verge of extinction (Mittermeier and Cheney, 1987). They are noted for their grace and agility high in the upper canopy of the forests, and although they are often referred to as semi-brachiators, their consistent use of an arm-over-arm manner of locomotion and their physical form should confer on them the rank of a true brachiator. Spider monkeys, with their long arms, long curved fingers and non-existent thumbs, have all of the physical adaptations for brachiation (see Figure 6.9 and 6.10). Their long tails, which are truly prehensile with fingerprint-like pads, serve them well in their acrobatic locomotor style and their suspensory foraging posture. From a distance, it is difficult to distinguish between the sexes because they are similar in size and coloring and the females have a pendulous clitoris which closely resembles the penis of the male (Napier and Napier, 1985).

Spider monkeys subsist mainly on a diet of fruit and exhibit the fission-fusion type of social pattern similar to that of chimpanzees and ruffed lemurs (*Varecia variagiata*), who also have a preference for fruit (Periera et al., 1988). The large groups of spider monkeys that congregate at night, break up into smaller foraging parties that range widely over a large area of the forest during the day. The composition of the small units can vary on a daily basis, and the same individual may be seen traveling with a variety of age/sex groups. It is suggested that this fluid social system would provide an adaptive edge, allowing the

Figure 6.9 A female spider monkey (*Ateles geoffroyi*). Note the long arms, long, curved fingers and prehensile tail. (Photo courtesy Linda Fedigan.)

animals to take full advantage of the patchy distribution of fruit and fluctuations in the amount available (Robinson and Janson, 1987).

Woolly Monkets and Woolly Spider Monkeys *(Genera Lagothrix and Brachyteles)*

The woolly monkeys (genus *Lagothrix*) are large New World primates with powerful prehensile tails. The fingertip-like pads that facilitate gripping cover about one-third of the undersurface of their tails, allowing them to hang indefinitely from the branches when feeding. Although it is suspected that they subsist on a diet primarily of friut, little more is known about their ecology and behavior in the wild. They are found in the middle and upper regions of the Amazon where habitat destruction has been very severe. Because they have been heavily hunted for food and require undisturbed forest for survival, *Lagothrix* is one of the most endangered New World genera (Mittermeier and Cheney, 1987).

The woolly spider monkey (*Brachyteles arachnoides*), is the largest New World monkey and has the dubious distinction of joining the golden lion tamarin (*Leontopithecus chrysomelas*) as the most endangered neo-tropical species. Its range is restricted to the rain forests of southeastern Brazil, where the near destruction of its habitat has left it with little breathing room.Unfortunately, to add to its problems, the wild populations have also been heavily hunted for food and pets (Mittermeier and Cheney, 1985). As yet, there are no groups in captivity, so the introduction of captive-bred animals into the wild is not an option for the woolly spider monkey. The muriqui, as it is commonly called, has become a symbol of the conservation movement in Brazil, and it is hoped that by educating the public to the importance of saving the primates, it will be able to escape extinction.

Clearly, the early bias against the study of New World primate species has dissipated, as every year more information is published. Aside from the fact that there is a sense of urgency in learning all there is to know about the many species that are threatened, we are now able to add to our fund of knowledge about arboreal species and to make some valid comparisons with the life-styles of some of the better-known Old World species.

Figure 6.10 A female Panamanian spider monkey with her infant (*Ateles geoffroyi panamensis*). (Photo by Russell A. Mittermeier, Conservation International.)

Chapter 7

The Old World Monkeys

Suborder - Anthropoidea Infraorder - Catarrhini Superfamily - Cercopithecoidea

The Infraorder Catarrhini, a large taxonomic group of primates characterized by downward directed nostrils, is divided into two superfamilies: the Cercopithecoidea, or Old World monkeys, and the Hominoidea, representing apes and humans. The Old World monkeys are extremely diversified in the number of species, the number of animals within these species and physical attributes. Unlike the New World monkeys, who are restricted to the tropical forest, Old World monkeys can be found in a wide variety of habitats with vastly different climates and vegetation types. There is only one family of Old World monkeys, the Cercopthecidae, which is composed of two distinct subfamilies: the Cercopithecinae and the Colobinae. Although these subfamilies have their own unique adaptations, they have the following features in common:

1. A diurnal lifestyle.
2. A gestation period of approximately 180 days, regardless of body size.
3. Single births.
4. The same **dental formula**, $\frac{2.1.2.3}{2.1.2.3}$,[1] which is also common to apes and humans.
5. **Ischial callosities**, or callous-like pads on the buttocks.
6. A bony tube connecting the ear drum to the outer ear.

The major differences between the cercopithecines and the colobines are related to dietary adaptations. The cercopithecines have cheek pouches, broader incisors and molar teeth with low cusps suitable for eating a diet composed at least partially of fruit. The colobines are basically folivorous monkeys with narrow incisors and high-cusped molars suitable for shredding and chewing leaves, and a large sacculated stomach adapted to digest large amounts of fibrous material. The colobines spend more of their time feeding and digesting their food than the cercopithecine species. The cheek pouches of the cercopithecines reduce the amount of time spent at the feeding site by increasing the amount to food that can be collected and eaten at a later date (Fleagle, 1988). The longer hind limbs, long fingers and shortened thumbs of the colobinae, especially evident in the colobus monkeys, are related to their enhanced ability to leap through the trees (Struhsaker, 1986).

As it would be impossible to discuss even a representative number of the approximately 80 species of Old World monkeys, the following chapter describes the major taxonomic groups, with a more detailed treatment of some of the better-known species.

Subfamily Cercopithecine

Except for the macaques, who live predominantly in Asia, the cercopithecines are an African group of monkeys. They outstrip the colobines in numbers of species and animals and in the diversity in their habitat use and physical features. There are nine genera of Cercopithecine monkeys, some of which include the species that have been studied since the early days of primatology.

Macaques (Genus Macaca) (Table 7.1)

The macaques are the most widely distributed of all of the nonhuman primates, with 17–19 species ranging from Gibralter in the west to Japan, Taiwan and Indonesia in the east. Their medium size and generalized limb structure, as well as their ability to co-exist with humans in many areas, has likely played a large role in their success. They are omnivorous and can be found in almost any ecological niche: tropical forests, mountain forests, open woodlands, towns and temples (Fleagle, 1988).

The behavior of macaques has been more heavily researched than any other primate form, primarily due to the diligence of Japanese primatologists who have kept continuous records of Japanese macaque groups since the early 1950s and the long-term studies of the rhesus macaques on Cayo Santiago. The general physiology of macaques is well known from the extensive work carried out on rhesus monkeys in biomedical laboratories. The diversity of climates and habitats in which they occur has resulted in differences in dietary, behavioral and physical adaptations. However, macaque species still have a number of characteristics in common. Most of the species live in large multi-male multi-female social groups from which males disperse. The females remain in their natal groups for life and build up large matrilines, while the males peripheralize at puberty and often join another troop or become solitary animals. Strong male and female dominance heirarchies play an important role in the social lives of most macaques species.

Japanese macaques (*Macaca fuscata*) deserve special mention because of the enormous body of data that has been collected by Japanese and Western primatologists, covering all aspects of their social and physical lives. This adaptable species is indigenous to the islands of Japan, and can be found as far north as the tip of Honshu at 41 degrees North latitude—farther north than any other nonhuman primate species. These monkeys have a unique place in Japanese mythology and are looked upon as a national treasure by the Japanese people. Japanese primatologists originated the idea of provisioning their free-ranging monkeys to habituate them to human presence which allowed researchers to identify specific animals and study their social system. Many important social concepts, such as kinship and dominance, have been advanced due to the interest of the Japanese researchers in the social patterns of their monkeys. Because of this fund of knowledge, Japanese macaque society will often be used to illustrate these concepts in Part 3 of this volume.

Table 7.1 The Infraorder Catarrhini - Family Cercopithecidae

Subfamily Cercopithecinae - Macaques	
Species	**Common Name**
Macaca fuscata	Japanese macaque
Macaca mulatta	Rhesus macaque
Macaca silenus	Lion-tailed macaque
Macaca tonkeana	Tonkean macaque
Macaca maura	Moor macaque
Macaca ochreata	Ochre macaque, Sulawesi booted macaque
Macaca brunescens	Muna-Butung macaque
Macaca hecki	Heck's macaque
Macaca nemestrina	Pig-tailed macaque
Macaca nigriscens	Gorontalo macaque
Macaca nigra	Celebes black macaque, Sulawesi crested macaque
Macaca sylvanus	Barbary macaque
Macaca sinica	Toque macaque
Macaca radiata	Bonnet macaque
Macaca assamensis	Assamese macaque
Macaca thibetana	Thibetan macaque
Macaca fascicularis	Crab-eating macaque
Macaca cyclopes	Taiwan macaque
Macaca arctoides	Bear macaque, Red-faced stump-tailed macaque

Sources: Adapted Napier and Napier, 1985 and from Fleagle, 1988.

Japanese macaques are greyish brown, medium-sized monkeys with short stubby tails (see Figure 7.1). Their multi-male multi-female groups range from 40 animals in unprovisioned groups to over 500 when the animals are fed artificially. They are opportunistic omnivores who will eat fruit, flowers, grasses, birds eggs and almost anything else available. Their adaptability in terms of diet and behavioral flexibility has been proven by the successful relocation of 150 monkeys from the mountain forests near Kyoto to the hot dry semi-desert area of south Texas. The Arashiyama West troop of Japanese monkeys has flourished in Texas since 1972, and now boasts well over 500 animals. The transplanted monkeys immediately began to forage and feed on the local vegetation (see Figure 7.2) and adjusted their daily activity patterns to accommodate the radical change in climate (Fedigan and Asquith, 1991).

The society of Japanese macaques follows the general pattern of other macaque species, with male dispersal and female philopatry, and rigid dominance hierarchies in

Figure 7.1 Japanese macaque male (*Macaca fuscata*). (Courtesy Linda Fedigan.)

both males and females. The **alpha,** or dominant, male and female are central to the group in terms of location and social activity. Kinship revolves around groups of related females and the large matrilineages found in Japanese macaque groups are made up of individuals descended from a common female ancestor. Japanese primatologists have noted that dominance structures are not only found in individual relationships, but extend to whole matrilineages, with the kinship group of the dominant female being dominant over all other **matrilineages.**

Mating season takes place in the fall, when the whole tenor of life changes from the relatively serene existence of foraging and socializing to one of frantic activity characterized by chases, vocalizations and sexual behavior. The physical appearance of the monkeys also changes in mating season, when the faces and sexual skin of the males and females turns bright red. Japanese macaques are among the few species of primates that are **series mounters,** where the male must mount the female several times consecutively in order for ejaculation to occur. This type of copulatory technique requires a great deal of cooperation between the mating pair and it is not surprising that Japanese macaques males and females form **consort bonds** during mating season, spending days or weeks interacting exclusively with each other (Fedigan, 1992)

Rhesus macaques (*Macaca mulatta*), who are very similar to the the Japanese monkeys in social patterns, sexual behavior and appearance, can be distinguished by their longer tails. Like Japanese macaques, they are extremely adaptable animals and can be found living in the temperate zones of the Himalayan Mountains and in tropical areas in the Indian subcontinent. These monkeys have survived not only in the wild in India but have

Figure 7.2 Young Japanese macaque of the Arashiyama West Group in South Texas, investigating a cactus plant. (Courtesy Linda Fedigan).

become "temple monkeys" and have flourished in urban areas because of their important place in the Hindu religion. India once exported over 200,000 rhesus monkeys per year to Western laboratories for use in biomedical research. However, this practice has been halted and now laboratory animals must be taken from breeding colonies (Peterson, 1989). The data from the Caribbean Primate Center in Cayo Santiago which has been collected since the late 1950s, as well as the extensive laboratory research on rhesus macaques, has provided us with a remarkable knowledge of their social and physical attributes as well as an insight into general primate characteristics.

Baboons (Genus Papio)(Table 7.2)

The controversy surrounding the taxonomy of the cercopithecines is especially evident in the classification of baboons. Some primatologists divide the genus *Papio* into five species, which includes four separate species of savannah baboons (*P. cynocephalus, P. usrinis, P. anubis* and *P. papio*) and the Hamadryas baboons (*P. hamadryas*), who live in the arid rocky areas of Ethiopia and Northern Africa (Napier and Napier, 1985; Fleagle, 1988). However, since the savannah-dwelling forms of baboon are so similar in social patterns and physical features, differing somewhat in size and coloring, it is less confusing to follow Thorington and Groves (1970), and treat the different forms as geographical races of *Papio cynocephalus*, the common baboon (see Figure 7.3).

Baboons are widely distributed throughout sub-Saharan Africa from Senegal to the Cape of Good Hope. They are the most conspicuous of the African primates and the ones that have come into the closest contact with humans, so it is not surprising that few primate genera have been more widely studied (Kingdon, 1983). The fossil record shows that early hominids were hunting baboons, indicating that the relationship with humans has been going on for about four million years.

Baboons are the largest of the monkeys, and the males, who are twice the size of the females, look particularly imposing with their heavy shoulder capes and lethal-looking

81

Table 7.2 The Infraorder Catarrhini - Family Cercopithicidae

Subfamily Cercopithecinae	
Species	**Common Name**
Savannah baboons	
Papio cynocephalus cynocephalus	Yellow baboon
Papio cynocephalus papio	Guinea baboon
Papio cynocephalus anubis	Olive baboon
Papio cynocephalus ursinus	Chacma baboon
Sources: Adapted from Thorrington and Groves, 1970.	
Papio hamadryas	Hamadryas baboon
Mandrillus sphinx	Mandrill
Mandrillus leucophaeus	Drill
Theropithecus gelada	Gelada

Sources: Adapted from Napier and Napier, 1985 and Fleagle, 1988.

Figure 7.3 Female savannah baboon with infant (*Papio cynocephalus*). (Photo by Brian Keating. Courtesy of the Calgary Zoological Society.)

canines. The species name *cynocephalus* is Latin for "dog head," referring to the dog-like muzzle of the baboon which is adapted to house the prominent anterior teeth, rather than to enhance the sense of smell. The males have a distinctly leonine appearance and the sight of a number of large baboons baring their canines and barking is usually enough to frighten away a potential predator. Although the canines of male baboons could provide formidable weapons in aggressive encounters with conspecifics, there is little evidence of wounding or killing in baboon society.

All species of savannah baboons (*Papio cynocephalus*) are semi-terrestrial and range during the day in forested areas and open country. At night they find safe sleeping sites in the trees or rocky outcrops. They are omnivorous and subsist on a wide variety of foods, including grasses, fruit, leaves, tubers, small mammals and even small crustaceans (Napier and Napier, 1985). Some groups of baboons in Kenya appear to be particularly fond of meat and have been reported to hunt small vertebrates on a regular basis (Strum, 1987). Although their large multi-male, multi-female social groups usually have extensive home ranges, the size of the range varies with the availability of food. The early field studies reported that baboons lived in a highly structured, male-dominated society. However, subsequent research has indicated that the females play an important role in troop life. Although dominance hierarchies are apparent in both the males and females, the male dominance structure is far less stable due to their tendency to transfer between troops. Females remain in one group for life and form the core of the baboon society. They often have one or two male "friends" who associate with them on a regular basis and protect their infants. It appears that a male's position in the group is dependent on his relationship with the females (Strum, 1987).

Although baboons do not have a discrete breeding season, there is a distinct birth peak just before the rainy season. Estrous females show very obvious red swellings of the sexual skin, and it has been documented that ovulation occurs at the height of the tumescence. Although the females mate with a number of males during their estrous cycles and the paternity of the infants is uncertain, baboon males tend to be protective of the infants of their female "friends." The infants depend on their mothers for about a year before they are fully weaned, at which time the adult males tend to watch over and protect the juvenile baboons.

Hamadryas baboons (*Papio hamadryas*) live mainly in the rocky gorges of Ethiopia and Arabia. Although they are somewhat smaller than the common baboon, and the cape of the male is silver rather than brown, they are very similar in appearance and show a similar degree of sexual dimorphism (see Figure 7.4). Hans Kummer, a noted Swiss primatologist who first studied hamadryas baboons in the 1960s, has provided much of our knowledge to date on their ecology and social structure.

Figure 7.4 A male and a female hamadryas baboon (*Papio hamadryas*). The male is twice the size of the female. (Courtesy Ruben Kaufman.)

Unlike savanna baboons, the basic social system of hamadryas baboons is a polygynous society that operates on three different levels of complexity:

1. The first level is composed of the polygynous groups and all-male groups. All of the social and reproductive activities take place in these small units. The resident male in the heterosexual groups is very intolerant of other adult males and zealously defends his "harem."[2] In contrast to almost every other primate society where males solicit females by affiliative gestures, the hamadryas male keeps his females near by herding them with threats and bites to the neck. As in other polygynous primate societies, the males in the all-male groups are closely bonded. Although the male hamadryas baboon is clearly dominant, the females are closely affiliated, and if the male is removed from the group, one of their ranks takes over the male role and the group continues to exist (Fedigan, 1992).

2. The next level in the hamadryas social system is commonly known as a "band," which is made up of of a number of smaller units that gather together to share safe sleeping sites on the cliffs. These are not random gathering of units, but are made up of animals that are familiar with each other and meet on a regular basis. The bands of hamadryas monkeys fission into their polygynous groups in the mornings and forage widely for dry leaves, beans and berries during the dry season. They prefer flowers and grasses in the wet season. In the evenings they return to the same sleeping cliffs to rejoin the other social units (Stammbach, 1987).

3. The third level is not a social group per se, but a large amorphous "herd," sometimes numbering 600 animals, who travel and forage together when food is abundant (Stammbach, 1987).

Kummer (1968) was able to identify two ways in which a male could establish his own polygynous group of females. A young male who was able to kidnap a juvenile female from a polygynous group would adopt her and care for her until she became sexually mature. At this point, the male would begin to herd her and she would become the first female in his reproductive group. In other cases, a follower male would stay close to the polygynous group waiting for his opportunity to replace the resident male.

The hamadryas reproductive parmeters, such as gestation lengths and interbirth intervals, are similar to those of the savannah baboons, and although there is no clear birth season, the estrous state of the females is evident from the large genital swellings.

Mandrills, Drills and Geladas (Genera Mandrillus and Theropithecus) (Table 7.2)

Some primatologists refer to mandrills and drills (Genus *Mandrillus*) and the geladas (Genus *Theropithecus*) as baboons (Stammbach, 1987; Fleagle, 1988). However, despite many similarities in appearance, biochemical and anatomical differences suggest that these species have been evolving separately from baboons for millions of years (Kingdon, 1974). Mandrills and drills are large, sexually dimorphic monkeys who live in the dense rain

Figure 7.5   A male mandrill (*Mandrillus sphinx*).
(Courtesy Linda Fedigan.)

forests of west Africa. The males are not only much larger than the females, but have colorful faces and rumps and pronounced ridges on their long snouts (see Figure 7.5). It has been suggested that the brightly colored faces and hind-quarters of the male mandrills serve as visual signals to facilitate group cohesion. Little is known about the behavior and ecology of *Mandrillus*, beyond the fact that they are primarily vegetarians who range widely in groups with one or two adult males and several females. These uni-male groups combine to form large herds of up to 250 animals in the dry season. The males appear to be highly terrestrial, foraging on the forest floor, while the females and young have been observed feeding in the trees (Stammbach, 1987).

Geladas (*Theropithecus gelada*) live among the cliffs and high plateaus of Ethiopia and are the most terrestrial of the nonhuman primates. They are large animals and, like baboons, the males are twice the size of the females, with heavy shoulder capes and extremely formidable canines. Geladas are often referred to as "bleeding heart monkeys" because both males and females have a patch of naked pink skin on their chests which is surrounded with bead-like vesicles in the females. In estrous females, the vesicles swell and the area becomes reddened, a pattern which is repeated in their sexual skin (Napier and Napier, 1985).

Although their common mode of locomotion is quadrupedal, geladas forage by shuffling along the mountain grasslands on their ischial callosities and fatty buttock pads for the grasses, seeds, roots and flowers that make up their diet. Their mobile fingers and opposable thumbs are particularly adapted for picking up small grass seeds, and their precision grip is superior to that of any other nonhuman primate.

The gelada social structure is similar to that of the hamadryas baboons, with the same three levels of groupings. The basic social units are the polygynous uni-male and the all-male groups, with the usual solitary, or follower, males. These units gather together in clusters, which are often called bands, to share safe sleeping sites on the cliffs. According to Dunbar and Dunbar (1975), the bands of geladas are quite variable and not as stable in composition as hamadryas bands. The third level of organization is the herd, which is not a true social group, but rather a large aggregation of animals who occasionally forage together.

Although the three-tiered social system of the geladas is similar to that of the hamadryas baboons, their patterns of social behavior are very different. Unlike hamadryas society, which is so clearly male-based, the dominant female gelada leads group movements and the male can be coerced by an alliance of closely-bonded females. The male appears to be unable to prevent follower males from mating with females who openly solicit copulations. The polygynous male's tenure in the social group rests with his ability to fight off the challenges of these males. It is surprising that these takeover attempts do not inflict severe injuries, given the lethal quality of the gelada canines (Dunbar and Dunbar, 1975).

Patas Monkey *(Erythrocebus patas)* *(Table 7.3)*

The patas monkeys (*Erythrocebus patas*) are close relatives of the forest-dwelling guenons, who have adapted to life on the hostile steppe-periphery of sub-Saharan Africa. They are widely distributed from Senegal to the Sudan and are basically a ground-dwelling primate like the gelada. Patas monkeys are medium-sized primates who show a striking degree of dimorphism in size and coloration. The males are approximately twice the size and more brightly colored than the females, and their handsome moustaches have prompted some to call them "the military monkeys" (see Figure 7.6). They are built on the

Figure 7.6   A male patas monkey (*Erythrocebus patas*) in a typical vigilant pose. (Photo by Pat McCosky. Courtesy of the Calgary Zoological Society.)

Table 7.3 The Infraorder Catarrhini - Family Cercopithecidae

Subfamily Cercopithecinae	
Species	**Common Name**
Erythrocebus patas	Patas monkey, military monkey, Hussar's monkey
Cercopithecus aethiops	Vervet, green monkey, grivet
Cercopithecus mitis	Blue monkey
Cercopithecus ascanius	Red-tailed monkey
Cercopithecus petaruista	Lesser spot-nosed monkey
Cercopithecus erythrogaster	Red-bellied monkey
Cercopithecus cephus	Moustached monkey
Cercopithecus erythrotis	Red-eared monkey
Cercopithecus mona	Mona monkey
Cercopithecus campbelli	Campbell's monkey
Cercopithecus pogonias	Crowned monkey
Cercopithecus denti	Dent's guenon
Cercopithecus wolfi	Wolf's monkey, Wolf's guenon
Cercopithecus diana	Diana monkey
Cercopithecus dryas	Dryas monkey
Cercopithecus salongo	Zaire diana monkey
Cercopithecus neglectus	De Brazza's monkey
Cercopithecus preussi	Preuss' monkey
Cercopithecus lhoesti	L'Hoest's monkey
Cercopithecus hamlyni	Hamlyn's monkey
Allenopithecus nigroviridis	Allen's swamp monkey
Miopithecus talapoin	Talapoin monkey

Sources: Adapted from Napier and Napier, 1985 and Fleagle, 1988.

order of a greyhound dog, with their long legs and slender bodies allowing them to move rapidly in the open country. They have been clocked at 55 kilometers per hour and have earned the reputation as the fastest of the nonhuman primates (Kingdon, 1974). Although they walk and run quadrupedally, they are quite adept at standing bipedally to survey their territory for potential predators. Their omnivorous diet consists of grasses, nuts, fruits and seeds supplemented with insects and lizards. Although they forage and feed on the ground, where they appear to be most comfortable, they usually sleep in the trees (Struhsaker and Gartlan, 1970).

They live in polygynous and all-male groups, but unlike the hamadryas baboons and geladas, they do not form larger units. The resident male in the heterosexual units is socially and spatially peripheral to the rest of the group and his main role is that of a look-out and a diversionary in case of predator attacks. The male displays vigorously when confronted by a predator, thereby deflecting attention from the females and young who either flee or freeze. The male is very intolerant of other adult males, and it is suspected that he is also watching for all-male groups. The competition for females is a constant threat, and it has been reported that tenure as the breeding male often lasts only a few months.

The females remain in their social groups for life and form the core of patas society. Despite the difference in size, the females readily coalesce against the male and rarely interact with him except during mating season. Allomothering plays an important role patas society, as adult and juvenile females will caretake infants at the second week of life, widening the social horizons of the infant beyond the mother at a very early age. The infants are small at birth, but develop rapidly compared to other Old World monkeys, and are usually fully weaned at six months of age. The females are sexually mature at an early age (2.5 years) and reproduce quickly, perhaps in response to the unpredictable environment. As an adaptation to life in an open and sometimes dangerous environment, patas monkeys blend well with their surroundings and keep their vocalizations to a minimum (Chism et. al., 1984).

Guenons (Genus Cercopithecus) (Table 7.3)

The guenons, who are the most common of the African monkeys, are distributed throughout sub-Saharan Africa. Although the majority of the approximately 20 species are arboreal and live in the forested areas of West Central Africa, the vervet (*Cercopithecus aethiops*), which is the most widespread species, is a semi-terrestrial, open-country form. Most species are omnivorous and subsist on a diet of all types of vegetable matter as well as some insects and small invertebrates (Napier and Napier, 1985).

They are generally medium-sized monkeys whose leaping ability and quadrupedal style of locomotion allows them to move well in dense tropical forests. The arboreal guenons are among the most brightly-colored primates, rivalling the tropical birds in their spectacular appearence (see Figure 7.7). In the Kibale and the Kakamega forests, where the greatest diversity of species occurs, the guenons can often be found foraging in large polyspecific groups (Cords, 1987). Kingdon (1974) speculates that the colorful faces and hind-quarters serve to differentiate the species to prevent cross-breeding, and also to communicate messages to conspecifics.

Despite the large number of guenon species, only a few have been studied on a long-term basis because of their dense habitat and their timidity resulting from persistent hunting. However, it has been established that most guenons live in polygynous social groups with one male and a number of females and their young. As is the case in most polygynous species, the females remain in their natal groups for life and the males leave at puberty, presumably to become solitary or to join and an all-male group. The social networks of most species are largely unknown because their interactions are subtle and infrequent compared to those of baboons and macaques. Reproductive parameters, such as gestation lengths, remain somewhat of a mystery, since none of the guenons, except for *Allenopithecus*, show any external signs of estrus (Cords, 1987).

Figure 7.7 A group of spectacularly colored Diana monkeys (*Cercophithecus diana*). (Photo by Brian Keating. Courtesy of the Calgary Zoological Society.)

The semi-terrestrial vervets (*Cercopithecus aethiops*) are the most widespread and successful of all the guenons. They are distributed from Senegal southward to the tip of South Africa and can even be found free-ranging in the Caribbean Islands of St.Kitts and Nevis, having been transported to the New World on slave ships. They have been called opportunistic omnivores because of their ability to eke out a living in areas where human disturbance has forced other primates species to leave. In regions where agriculture has replaced the forests, the vervets have become consummate crop-raiders and in areas of heavy tourism, they do very well on garbage and hand-outs. The behavioral variability that has been observed in regional vervet populations is likely a function of their ability to adapt to almost any kind of environmental upheaval. Where they are hunted as crop-raiders, they have developed a silent, vigilant behavioral pattern, while in protected parks like Amboseli Game Reserve, vervets have an amazing repertoire of vocalizations.

Vervets are relatively small for semi-terrestrial monkeys with only a moderate degree of sexual dimorphism (see Figure 7.8). Their grey bodies and black faces and hands contrast with their bright blue abdominal skin, which is particularly evident in the scrotum of the adult male. The long expressive tails of the vervets are often carried aloft during troop movements across the savannah (Fedigan and Fedigan, 1988). Unlike the forest-dwelling guenons, vervets live in large multi-male, multi-female groups that vary in size with food availability. There are clear dominance hierarchies in both males and females. However, as in many other multi-male social systems, the females dominance hierarchy is more stable because of the males' tendency to emigrate at sexual maturity. The females, who remain in their natal groups, form the core of the social group and readily unite against the males if necessary. The close bonds formed with members of their female kin represent an important aspect in the social lives of female vervets. **Allomothering** is widespread in vervet groups, likely because the mothers need the help of other females to raise their inordinately large infants. The young vervets are very **precocial** and mature rapidly (Cheney and Seyfarth, 1990). Many of the outward social interactions of vervets,

Figure 7.8   Vervet monkeys (*Cercopithecus aethiops*). (Courtesy Karen Dickey.)

including sexual behavior, are rapid and cryptic, which has given them a reputation in some areas as the "sexless" monkeys (Fedigan, 1992). However, Cheney and Seyfarth (1990) have reported an extensive repertoire of vocal communications. The chapter on communication in Part 3 will deal with complexities of vervet vocal messages in detail.

Talapoins *(Genus Miopithecus) (Table 7.3)*

Talapoins are the smallest of the cercopithecines, and although they are closely related to the guenons, they resemble the New World squirrel monkeys in their social patterns, habitat use and diet. They are arboreal monkeys who live in very large groups in the riverine rain forests of West Central Africa. There is a general tendency for large, terrestrial primates to live in larger groups and have wider home ranges than small-bodied animals. However, talapoins, like squirrel monkeys, have unusually large daily ranges, which is thought to be an adaptation to their heavy reliance on insects in their omnivorous diets (Gauthier-Hion, 1973). The talapoins also represent a puzzling deviation from the expected degree of social integration found in most cercopithecine societies. As in squirrel monkey society, most of the social activity takes place within age/sex classes, and the males and females travel in distinct subgroups, only interacting with each other during the three month mating season.

Talapoins are superb leapers and are noted for their ability to move rapidly through the forests. The larger groups of talapoins supplement their diets by raiding manioc crops, and have been observed escaping the farmers by leaping into nearby rivers and swimming away to safety (Napier and Napier, 1985).

Mangabeys *(Genus Cercocebus) (Table 7.4)*

Mangabeys are large, hollow-cheeked monkeys who live in the forests of Central Africa. Although all mangabeys are adapted to life in the tropical forests and are primarily frugivorous, the four species can be divided in two distinct types reflecting different styles of habitat use. The *C. torquatus* group are sexually dimorphic and spend a great deal of

Table 7.4 The Infraorder Catarrhini - Family Cercopithecidae

Subfamily Cercopithecinae - Mangabeys	
Species	**Common Name**
The *Cercopithecus torquatus* group	
Cercopithecis torquatus	White-collared mangabey, Sooty mangabey
Cercopithecus galaterus	Tana River mangabey, Golden-bellied mangabey
The *Cercopithecus albigena* group	
Cercopithecus albigena	Grey-cheeked mangabey, White-cheeked mangabey
Cercopithecus aterrimus	Black mangabey, Black crested mangabey

Sources: Adapted from Napier and Napier, 1985 and Fleagle, 1988.

time on the ground foraging through the leaf litter for grubs and mushrooms to supplement the fruit in their diet. The *C.albigena* group are arboreal, and like a number of tree-dwelling species, show little dimorphism. Some species live in uni-male groups, while other show a tendency to live in larger groups with a number of males (Napier and Napier, 1985). It has been suggested that mangabeys, who share the prominent estous swellings and grunt-like vocalizations of baboons, separated from them quite recently in evolutionary time (Kingdon, 1974).

Subfamily Colobinae

The Colobinae are much more exclusive and homogenous than the Cercopithecinae. The five to seven genera are all leaf-eating monkeys that are distributed throughout the tropical forests of Africa and a number of ecological zones of Asia. The main feature that distinguishes them from the cercopithecines is the large sacculated stomach with a fermentation chamber that allows them to digest cellulose and fibrous material as well as the toxins often found in mature leaves.

The members of the Colobinae belong to two major groups: the colobus monkeys of Africa and the langurs and odd-nosed monkeys of Asia. Many of the species have long fingers, shortened thumbs and unusually long feet which promotes a **saltatory** (leaping) motor pattern. Although most of the colobines are arboreal and restricted to life in the rain forests, the Hanuman (gray) langur is widely distributed in Asia in a number of ecological zones and is able to coexist with humans in urban areas of India. Many of the species are threatened due to widespread habitat destruction, primarily deforestation for agriculture and logging interests (Fleagle, 1988). The small bisexual unit is the most common social grouping of the colobines. However, the Hanuman langurs (*Presbytis entellus*) and the red colobus monkeys (*Colobus badius*) show variable social structures in different habitats. Two species of colobines (*Presbytis protenziani* and *Nasalis concolor*) live in monogamous family groups, which is extremely rare in Old World monkeys.

Colubus monkeys *(Genera Colobus) (Table 7.5)*

The species of colobus monkeys that make up the African branch of the Colobinae are closely related, but vary in coat color, social systems and behavior. The comparison between the tropical forest dwelling red colobus and the black and white colobus monkeys has been discussed in Part 1 to illustrate the difficulty in predicting social patterns by similarities in morphology, diet and habitat.

Table 7.5 The Infraorder Catarrhini - Family Cercopithecidae

Subfamily Colobinae - African Species	
Species	**Common Name**
Black and white colobus species	
Colobus guereza	Black and white colobus monkey
Colobus polykomos	King colobus monkey
Colubus angolensis	Angolian colobus monkey
Colobus satanas	Black colobus monkey
*Colobus badius** or *Piliocolobus badius*	Red colobus monkey
Procolobus verus	Olive colobus monkey

*Subspecies names not included
Sources: Adapted from Napier and Napier, 1985 and Fleagle, 1988.

The three species of black and white colobus monkeys are distributed throughout the rain forests of Sub-Saharan Africa from Senegal to Western Kenya. However, the type that has been studied most extensively is *Colobus guereza*. They are large, robust primates, who are highly prized by the natives for their luxurious black and white pelts. They are totally arboreal and live in small bisexual groups in the upper and middle canopy of the forest. It has been suggested that their small home ranges within the tropical forests are a function of their monotonous diet of the mature leaves from only a few tree species. Black and white colobus monkeys defend their territories vigorously against conspecifics and their loud, booming calls appear to function as a spacing mechanism. Infants, who are born with a neo-natal coat of pure white, are readily accessible to other females in the group. This pattern of allomothering, with the resulting social precocity of the young is seen to be an adaptation to promote social cohesion (see Figure 7.9) (Struhsaker and Oates, 1975).

The black colobus monkey (*Colobus satanus*) is one of the few seed-eating members of the Colobinae. Although they do eat some leaves, the primacy of hard seeds in their diet allows them to live in areas that are unavailable to other Colobus monkeys.

The red colobus monkeys (*Colobus badius*) are patchily distributed across equatorial Africa from Gambia to Zanzibar. Although they are all arboreal folivores, the enormous

Figure 7.9   Black and white colobus monkey
(*Colobus guereza*).

amount of interspecific variation in coat color and aspects of their social patterns has prompted some primatologists to consider the different forms as separate species, while others treat them as subspecies of *C.badius*. They are large, sexually dimorphic monkeys who live primarily in the upper canopy of the forests. Their locomotor pattern, while primarily quadrupedal, has been described by Struhsaker (1975) as "suicidal," referring to their impressive spread-eagled leaps into space.

Most of the types of red colobus monkeys live in the rain forests where they have been studied quite extensively (Struhsaker, 1975). However, two subspecies live in the woodland savannah and riverine forests of Gambia and Kenya, respectively. Although both forms subsist on young leaves and shoots, the diet of the rain forest monkeys is much more varied. The rain forest subspecies live in large, stable multi-male, multi-female groups who forage over a large daily range, while the other forms live in much smaller groups with only one or two adult males. The answer to the variability in the size of the social groups likely lies with differences in folivorous diet. Struhsaker (1975) reported that the males in the forest-dwelling groups form strong linear dominance hierarchies and show the tendency for group transfer seen in most multi-male societies. The females in these groups appear to be subordinate to the males in terms of access to resources and choice of mates. In contrast, both the males and females in the smaller groups are extremely mobile. The riverine and woodland savannah red colobus monkeys are among the few primate societies in which the females transfer more than the males. The females in these groups are active and aggressive, and seem to be much more successful in choosing their mates (Marsh, 1981). The conspicuous swelling of the sexual skin of the estrous females is not usually seen in colobines and is thought to be related to the multi-male social pattern of the red colobus monkeys. The infants who are born throughout the year are protected by their mothers from interactions with other group members, and take a relatively long time to mature socially compared to other members of the Colobinae.

Red colobus are reported to be relatively tame, unaggressive monkeys who show little territoriality and react with few low-intensity vocal displays when confronted by a group of conspecifics. The forest subspecies can usually be seen travelling in large polyspecific groups with black and white colobus and red-tailed monkeys (Struhsaker, 1975). Although the forest-dwelling species seem to be holding their own, the woodland savannah and riverine forest groups are not only suffering from loss of habitat but are heavily hunted for food.

The olive colubus monkey (*Procolobus verus*) is the smallest and one of the least well known of the colibine species. They share the folivorous diet, the tropical forest habitat, and the physical characteristics of the other colobus monkeys, and are said to display an extraordinarily versatile locomotor ability. Although the details of their reproductive paramaters are not known, the olive colobus resemble the red colobus monkeys in that the estrous females show genital swellings. The mothers carry the newborn infants around in their mouths, a maternal behavior that is unique in Anthropoidal species (Fleagle, 1988).

Langurs and Odd-nosed Monkeys (Genera *Presbytis, Nasalis, Pygathrix, Rhinopithecus* and *Simias*) (Tables 7.6 and 7.7)

These Asian leaf-eating monkeys show the greatest diversity and abundance of any of the Colobinae, especially in South and Eastern Asia, with the genus *Presbytis* being the most common. Most of the species are arboreal and live in polygynous societies. The Hanuman, or gray langurs (*Presbytis entellus*), who are the most terrestrial of the colobines, have proven to be an extremely adaptable species and are distributed from Sri Lanka to the Himalayan mountains in Nepal . They are important in the Hindu religious system, which credits the God Hanuman (a monkey) with sending his monkey troops to help the God Rama recover his kidnapped wife, thereby earning the eternal gratitude of the people. The langurs thrive in urban areas in India where they are are protected and fed, but seem to be able to survive equally well in rain forests, deciduous forests and deserts. They are large, greyish animals with pot bellies to accommodate the sacculated stomachs common to colobine species. They are primarily folivorous, but supplement their diet with fruit and flowers.

When primatologists who studied Hanuman langurs in the 1960s in northern and southern India reported completely different types of social patterns, it became clear that the environment had a significant influence on their social system. Jay (1965) found that the langurs in the sparsely populated areas of northern India lived in large, peaceful multi-male, multi-female groups, while Sugiyama (1964) noted that the south Indian

Table 7.6 Infraorder Catarrhini - Family Cercopithecidae

Subfamily Colobinae - Langurs	
Species	**Common Name**
Presbytis entellus	Hanuman langur, common langur, gray langur
Presbytis senex or *Presbytis vetulus*	Purple-faced monkey, Purple-faced leaf monkey
Presbytis johnii	Nilgiri leaf monkey, John's leaf monkey
Presbytis melalophos	Banded leaf monkey
Presbytis comata	Javan leaf monkey
Presbytis fontata	White-fronted leaf monkey
Presbytis hosei	Hose's leaf monkey
Presbytis aygula	Sunda Island leaf monkey
Presbytis rubicunda	Maroon leaf monkey
Presbytis thomasi	Thomas's leaf monkey
Presbytis potenziani	Mentawai leaf monkey
Presbytis cristata	Silvered leaf monkey
Presbytis francoisi	Francois' leaf monkey
Presbytis geei	Golden leaf monkey, Gee's leaf monkey
Presbytis obscura	Dusky leaf monkey
Presbytis phayrei	Phayre's leaf monkey
Presbytis pileata	Capped leaf monkey

Sources: Adapted from Napier and Napier, 1985 and Fleagle, 1988.

Table 7.7 The Infraorder Catarrhini - Family Cercopithecidae

Subfamily Colobinae - Odd Nosed Monkeys	
Species	**Common Name**
Nasalis larvatus	Proboscis monkey
Rhinopithecus roxellana	Golden snub-nosed monkey, Golden monkey
Rhinopithecus avunculus	Tonkin snub-nosed monkey
Rhinopithecus brelichi	Brelich's snub-nosed monkey
Pygathrix nemaeus	Douc langur
Simias concolor	Simakobu, Pig-tailed leaf monkey

Sources: Adapted from Napier and Napier, 1985 and Fleagle, 1988.

monkeys inhabiting crowded urban areas lived in uni-male groups. Moreover, the all-male groups that lived in conjunction with the uni-male groups were extremely aggressive and frequently attacked the bisexual units, killing the infants and attempting to drive out the resident male. At this point, the males competed against each other for access to the females who soon came into estrous after the loss of their infants. One explanation for this behavioral and social variability is that the high density population of humans and monkeys in southern India results in an increase in aggressive behavior. Another interpretation rests on the sociobiological theory that the males are acting to maximize their reproductive success by attempting to replace the resident males in the heterosexual groups and impregnate the estrous females (Hrdy, 1977 a and b).

Under normal circumstances the langur females have an infant every two years. They are extremely permissive mothers, and Jay (1985) reported that as many as eight females will take infants as far as 50 feet from their mothers on the first day of birth and caretake them for several hours.

The leaf monkeys (genus *Presbytis*) in southern and eastern Asia exceed all of the forest-dwelling vertebrates in both density and diversity. They vary widely in size and coloration, but are similar in physical conformation, with their flat faces, long limbs and tails, and protruberant stomachs. In most species, the females form the core of the social group, and are characterized by their affiliative behavior towards each other and the high level of allomothering. The tenure of the breeding male in the polygynous social group is usually short-lived due to frequent challenges from ouside males (Fleagle, 1988). For a detailed description of the physical features, ecology and behavior of the numerous species of leaf monkeys, the reader is directed to Primate Adaptation and Evolution by J.G. Fleagle (1988:183–189).

The Odd-Nosed Monkeys

Not surprisingly, the odd-nosed monkeys are a group of colibines characterized by their unusual nasal anatomy. The most well-known are the proboscis monkeys (*Genera nasalis*) who are large, red-coated primates easily identified by their pendulous noses. The males are twice as large as the females and, while the females' noses are large, the nasal appendages of the males are truly spectacular. It has been speculated that the adaptive value of such a large nose could lie in the production of the loud, honking alarm call of the adult males. Proboscis monkeys live in large closely-knit multi-male groups in the mangrove swamps of Borneo. Little is known of their ranging behaviors. However, they are adept swimmers and leapers and have little trouble moving through their watery habitat (Kanabe and Mano, 1972).

The golden, or snub-nosed, monkeys (Genera *Rhinopithecus*) of China and Viet Nam are the largest of the colobines, characterized by flat, bluish faces, snub-noses and thick luxuriant coats. One of the species, the golden monkeys (*Rhinopithecus roxellana*) live in the mountainous areas of China where they have to deal with long, snowy winters without the necessary physical adaptations. Their short, stocky legs coupled with the ability to move long distances to the mountain valleys in the winter indicates that they are basically semi-terrestrial primates. They live in large, multi-male groups of 20 to 30 animals in the winter, which appear to coalesce into large aggregations in the summer. Little is known about their ecology and behavior. However, since Chinese primatologists and conservationists are beginning to study these highly endangered primates, this situation is sure to be remedied (Fleagle, 1988).

The douc langur (*Pygathrix nemaeus*) is another interesting form of snub-nosed monkey. These large, brilliantly-colored monkeys are completely arboreal and live in small polygynous groups in the monsoon forests of Indo China. Although they appear to live on a vegetarian diet made up primarily of leaves in the wild, most of the information about their biology and social patterns comes from zoo colonies. The females play an important role in the group social activities, and as with most colobine species, allomothering of the newborn is common (Kavanagh, 1972). During the recent wars in Indochina, the douc langurs were widely hunted for food and the habitat destruction far exceeded the usual forms of forest exploitation. Since they rely entirely on the trees for their food, their continued existence in the wild is extremely uncertain (Peterson, 1989).

Summary

Although the Old World monkeys live in a diversity of habitats and are noted for their adaptability to all sorts of environmental conditions, they are remarkably uniform in morphology and behavior compared to the prosimians. Although they are often dimorphic, at least in the size of the canines, there are no really tiny cercopithecoids nor any inordinately large species. Their mode of locomotion is primarily quadrupedal running walking, either on the ground or in the trees, with some forms, particularly the colobines, who are especially good leapers. Most species are either omnivorous, with a preference for fruit, or folivores, with few species specializing on a diet of insects or gum like the small prosimians and New World monkeys. Even their social systems are primarily confined to single male or multi-male groups, in which the general trend is for males to transfer between groups and females to remain in their natal groups.

The success of the Old World monkeys in the numbers of species, numbers of animals and sheer biomass does not appear to be a function of their physical diversity, but is more likely related to their behavioral flexibility which allows them to exploit habitats that would otherwise be inaccessible to primates (Fleagle, 1988).

---

[1] The dental formula indicates the number of teeth of each type, listed from left to right as follows: incisors, canines, premolars and molars. The formula shows the number of teeth in one upper quadrant of the mouth listed on the upper line and one lower quadrant of the mouth listed below the line.

[2] Kummer coined the term "harem" to describe hamadry as society, since it was evident that the male had total control over the females in his group. This term was once used to describe other polygynous primate societies, but has since been dropped because it is not applicable to other polygynous groups.

Chapter 8

The Apes

Suborder - Anthropoidea   Infraorder - Catarrhini
Superfamily - Hominoidea

The five genera of modern hominoids are divided into three families: the Hylo-batidae (the lesser apes), the Pongidae (the great apes) and the Hominidae (humans). Clearly, the apes are not as diverse as the the Old World monkeys, the other member of the Infraorder Catarrhini, and not nearly as widespread, with only a few extant species compared to the many forms of monkeys. The smaller size and more generalized limb structure of the monkeys have allowed them to adapt to a much wider variety of environmental areas, while the apes are restricted to the tropical forests of Africa and southeast Asia (Fleagle, 1988).

Apes are anatomically adapted for brachiation and the use of suspensory postures, and exhibit a number of anatomical features that distinguish them from Old World monkeys. Whether or not they consistently use this mode of locomotion, their upright posture, long arms and long curved fingers are suited for brachiation. The elongated pelvic bones of the apes have the muscle attachments necessary to keep their internal organs in place when they are suspended from the branches. The fact that none of the apes have tails apparently does not hinder their ability to move through the tropical forests. The typical hominoid Y5 cusp pattern on the lower molars, the broadened palates and nasal regions and large brain cases are the main cranial and dental features that distinguish apes from Old World monkeys. All of the stages of development, from gestation and infant dependency to life span are prolonged in apes compared to those of all other nonhuman primates (Fleagle, 1988).

Family Hylobatidae - The Lesser Apes (See Table 8.1)

The Hylobatidae, or lesser apes, are the smallest and most successful of the ape forms in terms of numbers of animals surviving in the wild. There are five species of gibbons that average about 6.4kg (14 lbs) in weight and one species of the larger siamang that weighs about 10.5kg (23lbs) (Napier and Napier, 1985). Some consider the siamang as a separate genus (*Symphalangus*). However, since all of the species are so specialized in limb proportions and uniform in morphology, ecology, behavior and social patterns, they are frequently grouped together in the genus *Hylobates* (Richard, 1985). They are totally arboreal and live in the rain forests of Southeast Asia, ranging from mainland Burma and

Table 8.1 The Infraorder Catarrhini - Family Hylobatidae

Species	Common Name
Hylobates syndactylus	Siamang
Hylobates lar	Lar gibbon, White-handed gibbon
Hylobates concolor	Crested gibbon, black gibbon
Hylobates hoolock	Hoolock gibbon
Hylobates klossii	Kloss's gibbon, Mentawai Island gibbon
Hylobates agilis	Agile gibbon
Hylobates pileatus	Pileated gibbon, capped gibbon
Hylobates moloch	Silvery gibbon
Hylobates muelleri	Mueller's gibbon

Sources: Adapted from Napier and Napier, 1985 and Fleagle, 1988.

China to the islands of Borneo, Java and Sumatra. It is interesting to note that, except for the siamang (*Hylobates symphalangus*) and the white-handed gibbon (*Hylobates lar*), none of the other species share the same geographical area (see Figures 8.1 and 8.2).

The lesser apes are the aerial acrobats of the primate world and they are truly exciting to watch as they whirl arm over arm through the forests. They are also the most suspensory of the apes, as their typical feeding posture is hanging suspended from the branches by one arm. Their extremely long arms, hook-like fingers and flexible forelimb joints are typical of true brachiators. Unlike the great apes, they have ischial callosities which provide

Figure 8.1   A white-handed gibbon (*Hylobates lar*). Note the long arms and curved fingers that allow them to brachiate so gracefully through the trees. (Courtesy Linda Fedigan).

Figure 8.2 A beige male and a black female white-handed gibbon. Although the males and female show no sexual dimorphism in size, there is often a difference in coat color. (Photo by Brian Keating. Courtesy of the Calgary Zoological Society).

comfortable resting cushions when they sleep sitting in the trees with their long arms wound around their legs.

All of the lesser apes live in monogamous family groups consisting of a mating pair and two or three young. As in other monogamous species, there is little sexual dimorphism in body size or dental anatomy as the females are the same size as the males and share their dagger-like canines. Siamang males and females are always black; however, *Hylobates lar*, the best-studied species, are either black or buff, irrespective of the sex. All species are similar in appearance with their round, flat faces and large wide-spread eyes (Napier and Napier, 1985).

Except for the siamangs, who eat a significant amount of the leaves, the hylobatids are basically frugivorous. They are fiercely territorial and can be found daily defending their fruiting trees from other groups of conspecifics with spectacular aerial and vocal displays. Gibbons are noted for their loud, penetrating vocalizations which serve a variety of purposes. Adult males sing long solos just before daybreak, while a young males advertise their quest for a mate by calling loudly. A mated pair "sing" daily duets, which are specific to the couple and which become more intricate over the years. Gibbons are noted for their extreme territoriality and it is not uncommon for a group to have daily confrontations over fruiting trees. Females take an active part in the territorial displays and the female gibbon's vocal contribution is often referred to as a "great call." The high level of behavioral and social synchronicity, common to monogamous species, is also found in *Hylobates*.

Gibbons females do not exhibit sexual swellings during their estrous cycles, which only occur every two or more years. Single infants are born about every two years and in many species, particularly the siamang, the male takes an active part in caring for the young when they are about one year old. Unlike females in other primate species, gibbon females are fairly aloof and remain peripheral to the male and their young. When the juveniles reach about seven years of age, both parents become intolerant of them and they are forced to find mates of their own (Leighton, 1987). It has been noted that gibbon pairs will often allow a male offspring to set up a new social group near their fruiting trees (Tilson, 1981).

Although gibbons have few natural predators, continued deforestation for logging and agriculture will have a devastating effect on their populations, and unless wildlife reserves are set aside, their future is extremely uncertain (Leighton, 1987).

Family Pongidae - The Great Apes (Table 8.2)

There are four living species of great apes: the African chimpanzees (two species) gorillas, and the Asian orangutans. They are not only larger than the Hylobatidae, as their name suggests, but have the distinction of being the largest of the nonhuman primates. The great apes are tailless and have the upright trunks, long arms and curved fingers characteristic of brachiators but, unlike their smaller relatives, are not able to use this mode of locomotor on a regular basis because of their size. All are basically forest dwellers, although the chimpanzees are able to live in a wider variety of habitats. Despite the similarities in morphology and reproductive parameters, the three major ape forms live in very different social systems (Fleagle, 1988).

Table 8.2 The Infraorder Catarrhini - Family Pongidae

Species	Common Name
Pongo pygmaeus	Orangutan
Gorilla gorilla gorilla	Western lowland gorilla
Gorilla gorilla graueri	Eastern lowland gorilla
Gorilla gorilla beringei	Mountain gorilla
Pan troglodytes	Common chimpanzee
Pan paniscus	Pygmy chimpanzee, bonobo

Sources: Adapted from Napier and Napier, 1985 and Fleagle, 1988.

The Orangutan (Pongo pygmaeus)

The orangutans are the only great apes to live in Asia, and although they were once widely distributed throughout the continent, there are now only two subspecies inhabiting the islands of Borneo and Sumatra. Both types are large animals covered with shaggy red hair. However, the two forms can be easily distinguished. The Bornean males, with their solid bulk and huge pebbled cheek flaps resemble Buddahs (see Figure 8.3), while the Sumatran males, who are slimmer, with modest cheek flaps, moustaches and wispy beards have a mandarin-like appearance (see Figure 8.4). Sexual dimorphism is pronounced, as the males are twice the size of the females and have cheek flaps and capacious throat sacs.

Figure 8.3 A male Bornean orangutan (*Pongo pygmaeus*). Note the large, pebbled cheek flaps. (Photo by Brian Keating. Courtesy of the Calgary Zoological Society.)

Orangutans are the largest arboreal mammals, subsisting primarily on fruit, although their dentition and manual dexterity allow them to feed on other items such as bark, pith and insects. Their dental morphology is quite different from that of the other ape forms in that the upper central incisors are large and spatulate, and the lateral ones are small and peg-like. The thickened molar enamel resembles the human dental condition rather than that of chimpanzees and gorillas (Rodman and Mitani, 1987).

The limbs of the orangutan are specialized for suspensory postures, with very long forearms, mobile hip joints, prehensile hands and flexible feet with fully opposable toes. Because of their large size, orangutans move deliberately through the trees, using all four appendages to swing from branch to branch. Although the females and their young are almost totally arboreal, travelling in the middle canopy of the forest, the large males often

Figure 8.4 A male Sumatran orangutan (*Pongo pygmaeus*). (Photo by Brian Keating. Courtesy of the Calgary Zoological Society.)

have to cover part of their range on the ground, moving quadrupedally by supporting themselves on their fists. Despite their size, males and females spend the nights in sleeping nests high in the trees.

The orangutans are said to be the most solitary of all diurnal primates species, particularly the adult males who interact with the females only when they are in estrus. This social system (or lack of one) is said to be an adaptation to their frugivorous eating habits, since an adult male alone could quickly strip a fruiting tree. Like the nocturnal prosimians, the large ranges of the males overlap with those of several females. An adult male, who has established residency that encorporates the ranges of ovulating females, will form a consort bond with one of the estrous females and travel with her for several days. Females rarely travel together or interact socially, and the only formal social grouping in orangutan society is the matrifocal unit consisting of a female and her offspring. The adult males are very intolerant of other males and proclaim their territories with resounding "long" calls. They are dominant over subadult males and the long call seems to be sufficient warning to keep the younger animals away from the fruiting trees. Subadult males have been observed forcing their sexual attentions on unwilling anestrous females, a situation that is unknown in any other primate species. This type of liason never appears to result in a pregnancy (Rodman and Mitani, 1987).

Birute Galdikas, who was recruited by Louis Leakey and funded by National Geographic, has been studying orangutans in Borneo since 1971 and has contributed a great deal to our knowledge of their life-way patterns. Although she has observed only rare occurrences of tool use in the wild, the orangutans at her rehabilitation center in Tanjung Puting show an impressive ability to open locks and use sticks and kitchen utensils for a multitude of purposes. Although the scope of social interactions observed in wild orangutans is limited, the animals at the rehabilitation site are extremely gregarious (Galdikas, 1982). It would seem that exposure to humans releases a number of hidden talents in orangutans.

Given the lack of estrous swellings and the long interbirth intervals, details of the reproductive parameters of female orangutans, such as the age of first pregnancy, have been difficult to document. Orangutans are particularly slow breeders, due to the inordinately long interbirth interval (7 years) and the fact that females are not sexually mature until about 15 years of age. Clearly, the loss of even a single infant to a poacher represents a real threat to orangutan populations. Although the government of Indonesia is attempting to curtail the black market in pet orangutans and has promoted centers devoted to rehabilitating captive animals into the wild, widespread logging poses the greatest threat to the survival of the orangutans. It seems that only the maintainance of large national reserves and continued resistance to economic pressures will protect the orangutan from extinction in the wild (Rodman and Mitani, 1987).

The Gorilla (Gorilla gorilla)

There are three subspecies of gorillas living discontinuously in the tropical forest belt of Africa. The western lowland gorillas (G. g. gorilla) (see Figure 8.5), who are the most numerous, are found on the western side of the Congo basin, while the other subtypes, the eastern lowland gorillas (G. g. graueri) and the mountain gorillas (G. g. beringei) are

Figure 8.5   A male Western lowland gorilla (*Gorilla gorilla gorilla*). (Photo by Brian Keating. Courtesy of the Calgary Zoological Society.)

Figure 8.6   A male mountain gorilla (*Gorilla gorilla beringei*). (Photo by Brian Keating. Courtesy of the Calgary Zoological Society.)

restricted to small ranges in eastern Zaire and the Virunga Mountains on the Zaire/ Rwanda/Uganda border (see Figure 8.6). The gorilla is the largest of the modern apes with the adult males weighing approximately 155kg (350 lbs.) (Fleagle, 1988). Like the orangutan, gorillas are extremely sexually dimorphic in size and cranial features. The males are truly impressive animals, weighing twice as much as the females, with large **saggital**[1] and **nuchal**[2] crests, heavy brow ridges and longer canines. Gorillas have dense black or brownish-black fur covering their bodies and much of their face, giving them an inscrutable demeanor. The males, who develop a saddle of grey hairs on their backs when they reach 11 to 13 years of age, are referred to as **silverbacks.** These adult males are invariably dominant over the younger blackback males. Although they have all of the physical attributes for brachiation, adult gorillas are primarily ground-dwellers since there are few trees large enough to support their weight. Their usual style of locomotion is a sedate form of quadrupedalism, referred to as knuckle-walking. with the weight supported on their long curved fingers, The bipedal stance is only used by the males during their dramatic displays.

Although the mountain gorillas in the Virunga Volcanoes number only about 350 animals and are among the most endangered primate species, they have been studied intensively since the 1960s, first by George Schaller and then by Dian Fossey. Most of the information about gorilla ecology, behavior and social patterns has come from these studies (Stewart and Harcourt, 1987). Mountain gorillas are totally herbivorous and spend most of the daylight hours moving through the dense undergrowth, eating large quantities of leaves, pith, wild celery and other types of vegetation. At night, they use leaves and vines to prepare their sleeping nests on the ground. Since the distribution of the gorillas is severely limited by their need for great volumes of succulent greenery, they favor either dense forests or an open canopy with heavy ground growth. Not a great deal is known about the feeding ecology of the western lowland gorillas since they are difficult to find and habituate to the presence of researchers. However, there is evidence that their diet contains significant amounts of fruit when it is available. The group structure of the western lowland gorilla is the same as that of the mountain gorilla, and it appears that the dietary differences are only reflected in the larger daily ranges of the former subspecies (Tutin et al., 1992).

Gorilla groups, numbering from five to thirty animals, are composed of one silverback male, blackback males, females and their infants. The silverback is the only breeding male and Eisenberg et. al. (1972), who suggested that the other males are the silverback's sons, coined the term "age-graded society" to describe the gorilla social organization. Unlike the social pattern found in other multi- male, multi-female societies, the gorilla females are not bonded or even very affiliative, and often move out of their natal groups when solicited by solitary males or males from another group. The only severe aggression that has been observed in gorilla society occurs when outside males compete for the silverback's females. This situation triggers the branch throwing, chest beating displays so familiar to movie and television audiences. The silverback male, who is clearly the dominant animal, leads the group in their daily progressions and is the center of attention during the resting periods. Young males tend to leave their home group and become solitary males when they first achieve silverback status. Gorillas are among the least territorial of primates, as even when the core areas of different groups overlap, there appears to be little overt aggression (Stewart and Harcourt, 1987).

There is no definite birth season in gorilla society and the females, who show no visual signs of estrus, often solicit sexual behavior from the male. The single infants are very dependant when born and are not weaned until about 3 years of age. They are extremely susceptible to disease and injury and a female gorilla may only be able to produce three surviving infants over her lifetime. The adults are also prone to suffer from many human diseases and often succumb to arthritic disorders, pneumonia or heart trouble at an early age. Gorillas in captivity, who are not exposed to the rigors of life in the wild, can live to 40 or 50 years of age.

The negative attitude once held towards nonhuman primates was often directed towards gorillas, who were thought to kidnap and rape young women. Popular movies like *King Kong* did nothing to erase this image. These gentle animals did not deserve the reputation fostered by the media. However, their size and inscrutable expression likely added to the air of danger that surrounded them. Captive gorillas, who were often housed alone in featureless enclosures, became obese and catatonic and were thought to be stupid when compared to the gregarious chimpanzees. Laboratory studies have shown that gorillas are as adept at problem-solving and language acquisition as chimpanzees, given the proper environment. Fortunately, zoo conditions have improved greatly. Most facilities now take into account their physical and social needs, giving the public a chance to see gorillas with more normal behavior patterns. Some zoos have even been successful in raising captive-born infants.

Since all of the captive groups consist of lowland gorillas, successful breeding colonies will not help to bolster mountain gorilla populations. However, the plight of the mountain gorilla has become known internationally, largely due to the publicity given to Dian Fossey's research and her untimely death at her field site in the Rwanda. The fact that tourists will spend a great deal of money to see habituated gorilla groups high in the Virunga mountains has convinced the local governments of their value to the local economy. It is hoped that this situation continues, as the future of these impressive primates rests on the political stability of the African countries in which they live. Oko (1992) reported that, despite the efforts of CITES and the IPPL,[3] western lowland gorillas are still being poached for food and to supply the illegal international trade in infants (see Figure 8.7). Although there may be several thousand of these lowland gorillas in the wild at present, the poaching and the continued exploitation of the environment by timber and mining interests has placed them in jeopardy. Again it is up to the local governments to implement management policies and set up nature reserves if the gorillas are to be saved (Oko, 1992).

Figure 8.7 An infant mountain gorilla (*Gorilla gorilla beringei*). (Photo by Brian Keating. Courtesy of the Calgary Zoological Society.)

The Chimpanzee *(Genus Pan)*

Of all the great apes, the chimpanzees are the most closely related to humans, sharing 98.8% of our genetic material. Given the similarity in blood serum proteins and genetic makeup, as well as the resemblance in anatomical and behavioral characteristics, it is not surprising that chimpanzee society is often used as a model for early human social life. Although the common chimpanzee (*Pan troglodytes*) had been known to the scientific community since the 17th century, it was not until the 1930s that it was realized that there were two distinct types. In 1933, the pygmy chimpanzee (*Pan paniscus*) was classified as a separate species. At first glance, the two species appear very similar, with many of the same anatomical features, the same basic social organization and similar styles of habitat use.

However, major differences in behavior and the patterns of social interactions between the sexes make it worthwhile to discuss the species separately.

The common chimpanzees (*Pan troglodytes*) occur throughout the tropical rain forest belt in Africa, from Guinea to Uganda, Zaire and Tanzania (see Figure 8.8). Their relatively wide distribution, compared to the other species of apes, is likely related to their ability to survive not only in the tropical forests, but in the forest fringes and savannah woodlands (Kingdon, 1974). As their habitat disappears, they are gradually and relentlessly losing ground and are now listed in *The Red Data Book* of threatened and endangered species (Goodall, 1990). They are semi-terrestrial animals, who travel and socialize on the ground, but feed and rest in sleeping nests in the trees. Their long arms, short legs and curved fingers allow them to brachiate in the trees, but make knuckle-walking the only efficient style of quadrupedal locomotion on the ground. Although chimpanzees are basically omnivorous, eating nuts, leaves, honey, insects, monkeys and small ungulates, they prefer fruit when it is available.

Figure 8.8   A common chimpanzee (*Pan troglodytes*). (Photo by Brian Keating. Courtesy of the Calgary Zoological Society.)

Although they are powerful animals, chimpanzees are not nearly as large or as sexually dimorphic as the gorillas and orangutans. The males are somewhat heavier than the females and have longer canines and larger brow ridges. Like the gorillas, the chimpanzee coat is black. However, their bare faces are very expressive, making them appear less intimidating to human observers. Older animals, particularly females, are prone to baldness, and by our standards are much less attractive than young chimpanzees.

Although chimpanzees were studied by Nissen in the 1930s, much of our knowledge about their ecology, behaviors and social patterns has come from two study sites in Tanzania. Jane Goodall began observing the apes in Gombe Stream National Park in 1960 and has provided the world with an impressive body of longitudinal data on chimpanzee life-ways. In the 1970s, Gombe became a national research site sought out by students from all over the world. The contribution of Jane Goodall's efforts have been acknowledged not only in terms of her own research but also in terms of the noted primatologists who were inspired to continue in the discipline because of their introduction to field work in Gombe (Haraway, 1989). Japanese primatologists, working in the Mahale mountains south of Gombe since 1961, have also been the source of important ecological and social data.

Chimpanzee social structure has proven to be extremely complex. Goodall, in the early years of her study, observed small, temporary associations composed of a number of different age/sex groupings, and concluded that chimpanzees lacked a stable social unit. The Japanese team at Mahale mountain noticed, however, that there were large groups of chimpanzees who were familiar to each other and who greeted each other effusively when they met. These chimpanzee "communities," numbering from 30 to 80 individuals, were also found to occur at Gombe. The temporary groups observed by Goodall were foraging parties whose composition varied from day to day. These foraging parties were frequently **matrifocal** units, or all-male groups, and more rarely nursery groups of several females and their infants or a male consort with his estrous female. This fission-fusion type of social organization, which is reminiscent of that seen in spider monkeys, may have evolved as an adaptation to take advantage of the availability of fruit.

Wrangham (1979) has suggested that the ranges of the males are large and overlap those of several anestrous females, whose ranges may also overlap. Although males and female chimpanzees do not interact on a regular basis, females become attractive to males at the beginning of their estrous cycles when their perineal regions become red and swollen. There is no discrete breeding season and sexual behavior can be observed throughout the year. An estrous females may form a consort bond with one particular male or mate with a number of males during her cycle, making paternity difficult to establish. Young chimpanzees are not fully weaned until four years of age, and remain close to their mothers until they are about ten years old. Females often move to another community during their first or second estrous cycle while the males remain in their natal community. Male chimpanzees are extremely sociable and tend to travel and rest together, grooming each other frequently. In contrast, female chimpanzees live a relatively solitary existence, spending their days in the company of their infant and juvenile offspring. The male dominance hierarchy tends to be relatively unstable as the alpha male is constantly being challenged by individuals or alliances of subordinate males. The position of dominant male is only maintained by the judicious use of social strategies to coalesce support and is often bolstered by animated vegetation throwing and charging displays (Goodall, 1986).

Field studies (Goodall, 1986) have reported that chimpanzees show a rudimentary ability to use tools in the wild. Females search out suitable sticks and prepare them for the extraction of termites and ants, which they eat with evident relish. Leaves are chewed up and used as sponges to sop up the water, and appropriate stones and wooden clubs are transported to palm trees to crack open the nuts. Recently, there has been a great deal of interest in the healing plants eaten by the Tanzanian chimpanzees, particularly leaves containing chemicals that are effective against internal parasites (Turkington, 1985).

Chimpanzees are gregarious animals who use a variety of gestures and vocalizations for greetings, reunions, and reconciliations, and to communicate pleasure, grief, dominance, submission and affiliation. Some of their greetings, such as hugging, kissing and patting are very reminiscent of our own behaviors. Their "pant-hoots" can be heard ringing through the forest, advertising the presence of a large fruiting tree or merely adding emphasis to an exuberant display. Over the years, performing chimpanzees have been popular with circus audiences and their intelligence and willingness to learn has made them ideal subjects for ape language projects. Chimpanzees seem to enjoy learning human skills and performing before audiences. They become closely bonded to their trainers and are often treated like members of the family which, at first glance, does not seem to be an unpleasant fate for a young chimpanzee. However, an ethical problem arises when the cute, manageable young apes reach sexual maturity and become aggressive and less physically attractive. There does not seem to be a suitable place to retire these "over the hill" entertainers or research subjects, and they are often found languishing in small enclosures or in biomedical laboratories (Linden, 1985). Fortunately, there are caring primatologists, like Jane Goodall, who spend a great deal of time and effort raising the consciousness of the public to the plight of many captive and laboratory chimpanzees.

The distribution of the pygmy chimpanzee (*Pan paniscus*), commonly called the bonobo, is restricted to a small area of tropical swampy rain forest, south of the Zaire River. The term "pygmy" is inaccurate because the bonobo is not much smaller than the common chimpanzee, but has a slighter frame, narrower shoulders and a smaller, more rounded head. Although bonobos have the same basic patterns of habitat use as common chimpanzees, they are much more agile in the trees and their superb balance allows them to walk bipedally on a regular basis. There is a slight degree of sexual dimorphism in body weight. However, the dental and physical morphology of the males and the females is the same. Their omnivorous eating habits and preference for fruit are similar to those of the common chimpanzee (Fleagle, 1988).

Bonobos live in communities of 60 to 100 animals that break down into small foraging parties. However, the usual composition of these units and the patterns of social behaviors within the groups are very different from those of the common chimpanzee. In bonobos, the strongest affiliations are observed among adult females and between the sexes, and the most common foraging units are mixed parties which include adult males, adult females and their young and bisexual groupings of a male and female. Bonobo foraging parties are often composed of 16 or more animals, while common chimpanzees usually travel with no more than six individuals per unit. Aside from differences in the organization of the foraging parties, there are striking behavioral differences between the two species. Pygmy chimpanzee society is more peaceful and sociable than that of the common chimpanzee, with fewer **agonistic** encounters. The females, in particular, are more attentive to each other and show a number of affiliative actions such as grooming and food sharing which are rarely seen in common chimpanzee females. A genital-rubbing behavior, often observed between bonobo females about to compete for food, appears to act as a mechanism for reducing tension. In contrast, common chimpanzee females avoid tense situations by distancing themselves from each other. Bonobos also exhibit a greater degree of heterosexual activity and use a variety of precopulatory signals and copulatory positions. Despite the higher level of copulatory behavior and the fact that bonobo females are in estrous for longer periods, the interbirth intervals and general fecundity are the same for both species. It is suspected that the excessive sexual activity in bonobo society serves more to manipulate relationships than to enhance reproductive success. Although the females tend to leave their natal communities at their first or second estrous, they appear to have a very central position in bonobo society (Nishida and Hiraiwa-Hasagawa, 1987).

The deforestation of the ebony trees in their native habitat and the fact that they are often hunted for food has made the future of the pygmy chimpanzee doubtful. The instability of the Zaire government has forced researchers to abandon their study sites so that we may have difficulty adding to our meager knowledge of wild bonobos. Fortunately, primatologists are studying zoo colonies of pygmy chimpanzees, providing us with detailed information about their social behavior. Frans de Waal (1989) has documented the striking resemblance of captive bonobos to humans, in their behavior patterns and even in the games they play. The fact that bonobos, with their slender frames, effortless bipedalism and human-like behaviors, have almost replaced the common chimpanzee as the best model for early ape-like hominid, can only bode well for their future. We will almost certainly go to great lengths to assure the survival of a species that so closely resembles ourselves.

Summary

Although the four species of apes resemble each other in morphology and general ecological characteristics, the variability in their social systems is striking. All of the ape forms are diurnal, are restricted to forested areas close to fruiting trees and lush vegetation, and have the anatomical requirements for brachiation and suspensory postures. Their social patterns range from the monogamous family groups of the gibbons, the solitary orangutans, the age-graded societies of the gorillas to the fluid social system of the common chimpanzees and bonobos. Fleagle (1988) suggested that this social diversity could be related to a number of factors: 1) Apes are long lived and have time to adapt different reproductive strategies; 2) Their relatively large brains allow them the behavioral flexibility to form a variety of social relationships; or 3) Their size and relative invulnerability to predators reduces the constraints on their choice of social system.

As with the taxonomic categories of other primates, there are different opinions concerning the classification of the hominoids. The system of dividing the hominoids into the Hylobatidae, the Pongidae and the Hominidae used in this volume is based on morphological and behavioral similarities. There is, however, molecular evidence that the humans are closer genetically to the chimpanzees and gorillas than to the orangutans.

Therefore some primatologists group these apes with humans in the family Hominidae, and place the orangutans in the family Pongidae (Richard, 1985). Since the exact relationships between the great apes and humans are yet to be resolved, it would seem to be useful to continue using the traditional criteria of similar characteristics to classify the hominoids.

---

[1]A saggital orest is a bony ridge running along the midline of the skull that functions as an attachment area for the jaw muscles.

[2]A nuchal crest is a bony ridge at the back of the skull that serves as an attachment area for the neck muscles.

[3]CITES refers to the Washington Convention on International Trade of Endangered Species. IPPL refers to the International Primate Protection League.

Part III

Primatological Concepts

Chapter 9

Evolutionary Principles

Natural Selection versus Sexual Selection

Any discussion of evolutionary principles must begin with Darwin and his concept of evolution by **natural selection** (Darwin, 1859). According to Darwin, survival depends on the ability of an organism to reproduce successfully, and the most "fit" individuals are those who produce the most offspring that live to reproductive maturity. He observed that, while theoretically, every species has an unlimited reproductive potential, populations of animals remain relatively stable, so there must be a competition for survival. It was clear to Darwin that there was a great deal of individual variation within a species and that no two organisms were exactly alike. So, he surmised that the individuals with characteristics most suited to life in the environment would be the survivors. These favorable traits would be carried onto the following generations by the offspring of survivors and would eventually become predominant characteristics of that species. In short, the best-suited traits would be "naturally selected" to evolve.

Natural selection theory explains the evolution of traits that enhance the individual's struggle for survival. However, Darwin noticed that some animals possessed features, such as the male peacock's large tail, that seemed to work against survival. Since many of these traits were present in only one sex, Darwin (1871) proposed the **Theory of Sexual Selection** to explain the differences between the males and females of the same species. He used the term "secondary sexual characteristics" to describe those features, other than those necessary for reproduction, that were male/female specific, such as size, dental morphology, color, ornamentation and behavior. Numerous examples of primate species in which the males and females are different from each other in some respect were discussed in Part II, such as the large canines and cape of the male baboon, the saggital crest the male gorilla and the colorful face and hindquarters of the male mandrill (see Figure 9.1). In some cases, the physical characteristics are so extreme, as in the male peacock's tail, that it is difficult to see why they would have evolved. Sexual selection theory proposes that secondary sexual characteristics provide an edge in reproduction even though they do not directly act to improve the survival of the individual. The large canines of the male baboons could function to frighten away competitors for mates or to attract females. Darwin suggested that this reproductive advantage could come about in two ways:

1. **Epigamic** or **intersexual** selection - the interactions between the sexes. This is selection by preference which is generally discussed in terms of the female's

Figure 9.1   Sexual differentiation in mandrills. The male is twice the size of the female and has a brightly colored face. (Courtesy Linda Fedigan.)

choice of a male. Epigamic selection can explain the so called "useless" ornamentation of the male peacock as a feature to attract females.

2.   **Intrasexual** selection - the interactions between members of the same sex. This is selection by competition which is usually discussed in terms of the competition between males for access to estrous females. This principle of selection explains the evolution of the large-bodies and intimidating canines found the males of a number of primate species. The largest, most fearsome males would be expected to win the competitions for females and pass on the genetic basis for these traits to their offspring. Traditionally, intrasexual selection has been the principle that has received the most attention (Gray, 1985), although recent research into female sexual activity has shown that female choice has a very important role to play in the mating patterns of nonhuman primates.

The reason that these principles are not usually discussed in terms of male choice for females or female-female competition for mates relates directly to sex differences in reproductive potential. Males produce many low-cost sperm and theoretically can father an unlimited number of offspring. In contrast, females mammals must use a great deal of energy in bearing the infants, suckling and caring for them, and can only raise a limited number of offspring during their lifetimes. Thus males are selected to compete with other males for access to females and impregnate as many of them as possible, while females, with their limited reproductive capacity, would be selected to be extremely careful in their choice of their mates.

116

Sexual Selection and Sexual Dimorphism

Sexual dimorphism refers to the differences that exist between the males and females of a species. These include the differences in size, physical and dental morphology, coloration or behavior found in many of the species discussed in Part II. Although the degree of sexual dimorphism varies widely among primates, in most species the males are larger than the females. The explanation of sexual dimorphism in primates, primarily in body size, has been a major subject for theoretical discussion among evolutionary biologists.

At first glance, it might appear that the taxonomic affiliation of a species could be used to predict the occurrence of sex differences in size. Prosimians, New World monkeys and the lesser apes are usually monomorphic, while in some great apes and cercopithecine species of Old World monkeys the males are much larger than the females. However, the number of exceptions to this premise, especially in the case of Old World monkeys, makes taxonomic classification an unreliable predictor of sexual dimorphism.

Sexual dimorphism in primates may have evolved through the agency of sexual selection or by means of natural selection. To promote sexual selection as the primary causal agency, one would have to assume that the differences between the sexes resulted from differential mating success. To use natural selection as the explanation of sexual dimorphism, one would have to prove that the males and females differed in their use of the environment. Many of the explanations currently used to account for the evolution of sexual dimorphism in primates relate directly to the theory of sexual selection, especially the principle that deals with male-male competition. It is assumed that the largest males with the most frightening canines will win most of the competitions for females and father the most offspring. This, over time, favors selection for large males.

Mating patterns have been described as one of the factors that could determine the degree of sexual dimorphism in a given species of primate. Since there would be no need for competition in monogamous species where the male mates exclusively with one female, it could be predicted that the males and females would be the same size. In polygynous species, where the resident males in the heterosexual groups compete with outsider males, and in multi-male social groups, where there is competition for access to females, one would expect the males to be larger than the females. Although **monomorphism** is a characteristic of monogamous species, the evidence supporting this premise in polygynous and multi-male groups is not convincing. One of the problems in testing the validity of sexual selection theory as a major contributor to sexual dimorphism is the difficulty in assessing the degree of competition in primate groups. In large, open country species, where the competition from outsider males to take over the uni-male groups is obviously severe, the males are often twice the size of the females. In arboreal polygynous species, where the degree of sexual dimorphism is less extreme, there are no data to support the premise that the competition for the position of resident male is less severe. In some species that live in multi-male groups, like chimpanzees and many macaque species, there is very little sexual dimorphism in body size even though the competition for access to estrous females appears to be significant. In these cases, perhaps the competition is less rigorous that it appears. In order to assess the degree of competition between males, it is important to know how many males in polygynous and multi-male groups have access to copulations. It has been assumed that only one male in five or ten has access to females in polygynous species of animals, and therefore the competition is expected to be severe (Fedigan, 1992). In species like elephant seals, where only one-third of the bulls have an opportunity to mate, the competition is extremely fierce and the losers are often mortally wounded (leBoeuf, 1974). While this assumption may be valid for many mammalian species, it does not seem to hold true for primates. Field studies have shown that the males in polygynous groups are extremely mobile, joining and leaving groups as often as every four years therefore, most adult males will live in a breeding group at some time during their lives. The breeding success of males in multi-male groups and the actual degree of competition for females will remain a subject for debate until there is an accurate way of assessing biological paternity (Fedigan, 1992)

The fact that there is more sexual dimorphism in large-bodied forms such as baboons, geladas, gorillas and orangutans compared to the small-bodied species has often been attributed to the stronger male-male competition in the larger species. The socionomic ratio of males to females in a group has been cited as another factor affecting the amount of male-male competition and the degree of sexual dimorphism. Research has shown that in species where the females outnumber the males, there is a tendency for the males to be larger relative to the size of the females (Clutton Brock et al., 1977). The problem with using the socionomic ratio between the sexes to predict the degree of sexual dimorphism is that the relationship could be a function of overall body size. Large-bodied primates tend to be sexually dimorphic and often live in uni or multi-male groups, while in smaller-bodied species the males and females are similar in size.

To cite the sexual selection theory as the main factor contributing to sexual dimorphism is to assume that most of the size differences *cannot* be explained by natural selection (Gray, 1985). The following are some factors, other than male-male competition and mating success, that could account for the variability in size between the sexes:

1. Male Role

Males in most primate societies protect the females and the young from predators and thereby experience more predator pressure than females could select for larger-bodied males. Supporters of this explanation point to the evident sexual-dimorphism in terrestrial, open-country forms where presumably the threat from predators is greater than for tree-dwelling primates. Arboreal species are preyed upon by raptors and snakes rather than carnivores, and there is no real proof that ground-dwelling species suffer from more predator pressure than those who live high in the forest canopy.

2. Habitat Use

An alternate explanation for sexual dimorphism has been the exploitation of different ecological niches by the males and females of a particular species. Although this type of research has not been fully pursued, it has been observed that the males and females in some highly dimorphic primate species occupy different niches within their habitat, leading them to use different foraging strategies and to eat different foods. In orangutans, where the males are twice the size of the females, the males tend to travel and feed on the ground, while the females eat more food items found only in the trees (Galdikas, 1979). Similarly, female mandrills move and feed in the forest canopy, while the larger males forage on the ground for plants and insects (Jouventin, 1975). The validity of this explanation suffers from the possibility that the evolution of the differential in sizes between males and females preceded the exploitation of different niches.

3. Biological Function

In strongly sexually dimorphic species, the males and females are not only different in size but also have very different biological functions. The females bear the offspring and are the sole source of nourishment during the early post-natal period, regardless of whether or not they take the full initiative in the care and raising the infants. Since primate females spend most of their adult lives either pregnant or lactating, both of which require a great deal of energy, the energetic needs of the females may exceed those of the males. Although it was once thought that the larger males required more food than than females, research has shown that the females in a number of species eat more, spend more time foraging and feeding and are generally more active than the males (Clutton-Brock, 1977). If the energetic requirements are higher for females than for the larger-bodied males, it may be adaptively advantageous for the females to be smaller in order to reduce the amount of food needed for sustenance. Small body-size in females may have evolved to compensate for the costs of reproduction (Demment, 1984). The size differential would not be as extreme in small-bodied species, like many prosimians and New World monkeys, because the females would have to achieve a minimum size in order to bear the infants (Fedigan, 1992).

Summary

Many primatologists would agree that sexual selection is a major contributor to sexual dimorphism, at least in terms of size differences between males and females. The "selection by competition" principle has taken precedence over "selection by preference" in these discussions of male/female differences in body size (Gray, 1985). Sexual selection through male competition has been cited as a factor in determining the degree of sexual dimorphism through differences in mating patterns, the socionomic ratio between the sexes, ecological niche (open country vs arboreal habitat) and body size. Clearly, all of these variables are related because the large-bodied species that live in polygynous groups in the open country are almost always sexually dimorphic, and it is extremely difficult to tease out the input of each. It is also difficult to rule out natural selection as an important contributor to the evolution of sexual dimorphism. Factors such as the differences in the roles of the males and females in the society, differences in diet and habitat use, and the diversity in the energetic needs of the sexes could just as easily account for sexual dimorphism in body size. Gray (1985:208) summed up the lack of unanimity in the quest for a satisfactory explanation for the evolution on sexual dimorphism in primates when he wrote "In the case of sexual dimorphism of body morphology, it appears that most theorists would agree that while sexual selection is probably implicated in the evolution of body size dimorphism, it is not the only variable implicated, and its strength relative to the other variables is still an open research question."

Sociobiology

Natural selection, as defined by Darwin (1859), was the mechanism by which the fittest individuals survived to reproduce because they possessed the traits best suited for life in the environment. The key to fitness was survival. Sociobiological theorists view fitness in terms of reproductive success, where the the fittest individuals produce the most offspring and pass on more of their genetic material to the succeeding generations. Sociobiology is a relatively new discipline, coming to the forefront in the 1970s. E.O.Wilson (1975:4), who wrote the landmark book on the topic, defined sociobiology as "the systematic study of the biological basis of all social behavior." Sociobiologists argue that behavior has a genetic basis and that social behavior evolved by means of natural selection. The common thread that links sociobiological theories is the premise that all organisms are continually striving to maximize their reproductive success. The major tenet of the discipline is that individuals are always acting to insure the replication of their genetic material to the succeeding generation, no matter what outward appearances suggest.

The tendency of sociobiologists to use metaphors in their discussions has caused some confusion in the uninitiated. Where Darwin (1859) referred to environmental agencies such as disease or predators acting on survival and weeding out the less fit, sociobiologists describe individuals as actors whose goals are to maximize their own fitness in terms of reproductive success. Animals are conceptualized as strategists, making investments of time and energy in their reproductive output, while attempting to insure that the benefits outweigh the costs. It is important to remember that this terminology is only the sociobiologists' way of defining their principles and that the animals are not consciously figuring out behavioral strategies.

The agencies that produce behavior can be examined from three levels of causation, depending on one's point of view: the **ultimate** or evolutionary level, the proximate or immediate level, and the **ontogenetic** or developmental level. Sociobiologists only look at the ultimate causes of behavior that predict its adaptive value, and may tend to ignore equally valid proximate and ontogenetic causes. As an example of this bias, a sciobiologist would tend to explain the branch shaking display of a male Japanese macaque from an evolutionary point of view, arguing that it could have evolved to enhance reproductive success by attracting mates or maintaining a dominant rank. A proximate explanation of the same display would be that, since these behaviors escalate during mating season, they result from an extra secretion of male hormone. The ontogenetic point of view would explain the display as a behavior pattern that developed in the young as an attention-getting device to receive help or added care (Gray, 1985). Clearly, the three points of view are not necessarily mutually exclusive. Despite the evolutionary bias, one of the positive influences of the sociobiological approach is that it has stimulated researchers to go out into the field to study animals in their natural context. In turn, this has generated provocative questions about the adaptive function of behavior.

There are three major sociobiological principles that have pervasively influenced the field of animal behavior and which are are widely accepted by zoologists and primatologists. The following sociobiological theories discuss selection at the level of the individual or the gene, since it is the individual organism that is surviving and reproducing, and the gene that is the unit that is being inherited.

Kin Selection Theory

Over the years, researchers have puzzled over the fact that animals can be observed helping other individuals while putting themselves at risk and jeopardizing their survival and fitness in terms of reproductive success. The theory of **kin selection** was articulated by Hamilton (1963) to explain the evolution of this **altruistic**[1] behavior. The basic premise of Hamilton's theory is that since an individual's genetic material is shared by all of his or her biological relations in varying degrees, it is possible to have one's genes represented in succeeding generations without having reproducing. Individuals may help their kin, appearing to be self-sacrificing, when in reality they are promoting the replication of their genetic material to future generations through their biological relations,

Inclusive fitness is an important concept in kin selection theory. The term **individual fitness** refers to the ability of the individual to produce viable offspring and directly transfer genes to the next generation. Inclusive fitness is the "sum of an individual's own fitness plus the sum of the fitness of the relatives in proportion to their relatedness to that individual" (Drickamer and Vessey, 1982:67). The formula that Hamilton (1963) developed to predict the incidence of altruistic behavior includes three important factors:

1. The degree of relatedness between the two individuals (r);
2. The cost, in terms of the individual fitness of the donor of the altruistic behavior (C);
3. The benefit, in terms of the individual fitness of the recipient of the behavior (B).

The formula states that altruistic behavior can be predicted when the ratio of the recipient's benefits to the costs to the donor are greater than the reciprocal of the degree of relatedness (B/C = 1/r). In the case of brothers, who share one-half of their genetic material, the benefits would have to equal to twice the costs in order for altruistic to occur. Whereas with cousins, where the degree of relatedness is one-eighth, altruistic behavior would not be predicted unless the benefits equalled eight times the cost. The formula simply states that the expectation of altruistic behavior increases as the degree of relatedness increases, and so one would predict that more altruistic behavior would be directed towards close kin than to distant relatives.

The most extensive field study to test kin selection theory was carried out by Kurland (1977) on the Kaminyu group of Japanese macaques. He tested for relationships between the frequencies of grooming and defense behaviors (examples of altruism), and the degree of relatedness between animals. His results showed that the incidence of grooming and defence behaviors increased significantly with an increase in the degree of relatedness, and that the these behaviors were directed more frequently to kin than to non-kin. Although the findings appeared to support kin selection, Kurland did not have any adequate way of measuring the costs and benefits of grooming to individual fitness. He argued that the act of grooming could benefit individual fitness by ridding the recipient of external parasites and enhance its general health, while a cost to the groomer could involve internal problems from ingested hair. The definition of kinship between interacting animals was also ambiguous because paternity in Japanese macaques cannot be established and relationships can only be defined through the females.

Another problem that has plagued this study and others that have attempted to test kin selection theory is that there is an equally plausible proximate explanation for the higher incidence of helping behaviors to close kin. In Japanese macaque society, members of matrilineages grow up together and the females, in particular, spend most of their time in each other's company. The familiarity of related individuals with each other would almost certainly have an effect on the amount of grooming and defense behaviors shown towards them (Fedigan, 1992). Although the results of studies that support the predictions of the kin selection theory are not at all conclusive, it appears to be widely accepted among animal behaviorists, likely because of its elegance and simplicity (Gray, 1985).

Reciprocal Altruism Theory

While the theory of kin selection explains altruistic behavior between relatives, animals have been observed helping unrelated individuals. In this case, the loss of individual fitness is not offset by the indirect infusion of genetic material of kin to the next generation. The theory of reciprocal altruism, developed by Trivers (1971), stated that an individual would act unselfishly towards an unrelated conspecifics or a member of another species if it expected a return of the favor at a later date. Trivers cited warning calls in birds and the actions of cleanser fish to remove parasites from other fish species as examples of this type of altruism. In order for reciprocal altruism to be effective in a species, certain conditions must be met:

1. The species must be long-lived to allow time for the reciprocal act;
2. The individuals must meet repeatedly;
3. The organisms must be capable of learning and memory;
4. The benefits to the recipient must be large in comparison to the cost to the donor (Trivers, 1982).

Another proviso that is critical to the theory is that cheaters, or those who fail to repay the favor, must be recognized and excluded from the system.

Although it is difficult set up a project to provide empirical proof of reciprocal altruism, Packer's (1977) study of anubis baboons was designed to test the theory. He studied coalitions of adult male baboons and found that unrelated males had preferred partners who consistently assisted each other in fights and in solicitations of females. Although these results tended to support the theory, Packer had difficulty determining whether the benefits of the altruistic behavior exceeded the costs, and if cheaters were eliminated from the system (Gray, 1985). Barbara Smuts (1985), in a more recent study of savanna baboons, reported that older males formed alliances allowing them to overpower higher-ranking rivals for mates. Seemingly, these alliances were based on reciprocity, since males repeatedly solicited help from the same individuals and alternated the roles of helper and recipient. These alliances allowed the older males to have greater access to estrous females which would presumably improve their reproductive success. Although these results indicated that reciprocal altruism was operating in her baboon group, Smuts did not attempt to test the theory by looking at the costs versus the benefits of the alliances or the recognition of cheaters.

Parental Investment Theory

Classical Darwinian theorists and sociobiologists view the parent-infant relationship in different ways. While Darwin (1859) argued that the hostile forces of nature acted to eliminate the offspring with the least fit traits, sociobiologists focus on the efforts of the parents that are directed towards promoting the survival of the infant. The **parental investment theory**, which was also developed by Trivers (1972), states that "animals have been selected to invest their resources in a manner allowing them to achieve maximum inclusive fitness." (Gray, 1985:142). Parental investment is viewed as anything that is done to increase the current infant's chance of survival while decreasing the parents' ability to produce additional offspring. This investment includes the production of sex cells and any other time and energy expended on behalf of the infant (Trivers, 1972).

Parental investment theory emphasizes sex differences in the degree of parental investment in mammalian species. The considerable investment of the female includes the costs of the egg as well as the energy expended in bearing, nursing and raising the offspring, while the male's contribution, consists of only one low-cost sperm. The theory of sexual selection predicts this sex differential in parental investment. The females, whose reproductive potential is limited, would be selected to choosing a mate with good behavioral or genetic qualities, and then spend time and effort raising her offspring. The males, who produce many low-cost sperm, theoretically have unlimited reproductive potential. After competing with other males for access to females and mating with as many of them as possible, their best strategy would be to avoid expending energy on raising an infant who may not be theirs. In monogamous groups, where the male and female mate exclusively with each other, there is proof of paternity and the male would presumably be selected to take some part in the raising the offspring (Trivers, 1972).

Although to date, no studies have been designed specifically to test the parental investment theory, supporters of the theory point to the results of field studies that match the predictions to the observed behaviors. Problems arise in identifying the behaviors that fit the definition of parental investment and in assessing the costs to the parents versus the benefits to the current infant in terms of inclusive fitness. In most monogamous primate species, particularly the marmosets and tamarins, the males spend significant time and energy in raising the young. However, the evidence for male parental investment is less convincing in the case of polygynous and multi-male groups. In polygynous species, where there is only one breeding male in the social group, there are few examples of male care, other than as protector of the group. However, in multi-male groups where there is the least proof of paternity, there are some striking examples of male infant care. In Barbary macaques, adult males have been observed to spend a significant amount of time in the care and raising of the infants, while adult male savannah baboons will often groom and protect the infants of their female "friends" (Gray, 1985; Smuts, 1985).

The **parent-infant conflict** theory is an extension of the parental investment theory which was presented by Trivers (1974) to explain a situation, such as weaning, where there is an obvious clash between the parents and the offspring. He argued that the course of parental behavior can be predicted by viewing it as a function of the cost to the parents in terms of their reproductive success and the benefits to the survival of the current infant. In the case of weaning, the newborn infant will not survive without maternal care. Therefore, the investment on the part of the mother is advantageous to both parties. However, as the infant matures socially and physically, and is able to forage and feed on its own, the time comes when the mother is motivated to mate and have another infant while the infant still requires the comfort of its mother's nipple. At this point, there is a genetic tug-o-war between the mother and the infant, as their interests are now completely divergent, and the mother begins to reject her offspring. The infant, who is not a passive participant in the weaning process, acts to secure more maternal care than the mother is willing to provide by protesting with vocalizations and tantrums (Trivers, 1974). In the long run, the parent wins the conflict, because the infant must become independent and the parents must continue reproducing in order for parental behavior to evolve (Alexander, 1974).

Studies that have tested the parent-infant conflict theory on Japanese macaques have found that the frequency of maternal rejections increased significantly when the females were in estrus (Collinge, 1987, 1991; Worlein et al., 1988). Although there are valid proximate reasons why a female would not want an infant on her nipple while mating, this result does tend to support Trivers' (1974) theory that the mother would escalate her attempts to free herself of the current infant when she is motivated to mate and produce another infant.

Summary

The discipline of sociobiology is committed to viewing all social behavior in an evolutionary perspective. This has benefitted the field of animal behavior studies in that it has forced scientists to go into the field to study animals in their natural habitat and ask important questions about the adaptive significance of behavior. One of the criticisms of the discipline is that some sociobiologists have tended to overuse the concept of adaptation as the explanation of all behavior and have neglected equally compelling proximate explanations for animals behavior. Whether one approves or not, sociobiology has opened up a new way of looking at animal behavior that has stimulated discussion and expanded the horizons of field work. That, after all, is what science is all about.

---

[1]Altruistic behaviors are those behaviors directed towards an individual that increase its chance of survival at the expense of another's survival or reproductive potential.

Chapter 10

Socio-sexual Behavior

Introduction

The most striking aspect of sexual behavior in nonhuman primates is the enormous degree of variation observed in the different species. In vervets, whose sexual behaviors are very low key, the fleeting courtship gestures and brief copulations have earned them the reputation as the "sexless" monkeys. In Japanese macaques, the whole tenor of group life becomes directed towards sexual activity during the breeding season. Males chase females, females solicit males and the young practice mating skills (Fedigan, 1992). In pygmy chimpanzees, who have no breeding season, sexual activity takes place nearly every day throughout the year, often triggered by the excitement of sharing food (deWaal, 1989).

In 1932, Zuckerman published The Social Life of Monkeys and Apes, in which he stated that primates were continuously sexually active and that the sexual bond was the cement that held primate society together. Although this may be true for a number of animals species, it is not the case in nonhuman primate society where it has been established that the mother-infant bond is of primary importance. Zuckerman's argument was based partly on his observations of a group of hamadryas baboons at the Regent Park Zoo in London. His information was based on a highly unnatural social situation, where the inappropriate socionomic ratio of males to females triggered exaggerated levels of aggression and sexual activity.

Field studies in the 1950s and 1960s reported that infants in several species were being born at roughly the same time of year and that some species appeared to have a discrete mating season. At about the same time, laboratory studies were demonstrating the physiological phenomenon of "heat," periods of time when females were receptive to mating. It became clear that sexual activity was not continuous in many mammalian species, but occurred periodically in response to hormonal changes (Fedigan, 1992). It was found that in most primate groups, a female will likely spend only a total of about 20 weeks in mating activity over a lifetime of 20 years (Jolly, 1972). This being the case, it is unlikely that the sexual bond is the dominant factor in determining the stability of nonhuman primate societies.

Because stress affects the cyclicity in females and captive animals in zoos and laboratories quickly lose their seasonality, the phenomenon of mating seasons in nonhuman primates was only detected when primatologists began actively working in the field (Fedigan, 1992). Two main factors which could affect the hormonal state of the adult males and females and contribute to seasonality in animals are fluctuations in climate and photoperiod (hours of daylight). Most species live in areas where annual shifts in in temperature and/or rainfall affect the amount of food available. In tropical zones, where most primates live, rainfall fluctuation is the most common environmental variable. Because of the long delay between conception and birth in primates, environmental cues that forecast optimal conditions for birth in terms of the food availability may act to stimulate **endocrine** activity. Both males and females in seasonal breeding species experience changes in their hormonal output related to reproduction. The males generally only produce sperm at breeding season and the females experience estrus cycles, during which they are behaviorally motivated to mate. Rhesus and Japanese macaques, who have been studied under different climatic conditions, have shown some variation in the birth peaks, but continue to experience breeding season in the fall, so perhaps environmental agencies are not the only factors operating to stimulate sexual activity (Fedigan, 1992; Richard, 1985). It appears that the females are the key to the synchronicity in the physiological basis for the mating behaviors between the sexes. Studies have shown that the presence of an sexually-receptive female will bring about endocrine changes and mating behavior in a male, while a non-cycling female is not affected physically by proximity to a sexually-active male (Rose et al., 1972; Vandenberg & Post, 1976). Synchronicity in the estrus cycles of the females during breeding season is also important. It has been suggested that chemical communicators, known as **primer pheromones**, given off by the females are responsible for the coordination of the estrus cycles as well as the endocrine changes in the males.

Cyclicity and Estrus in Nonhuman Primates

The concept of estrus corresponds to the idea of "heat" in animal species and was once defined as a period around the time of ovulation when females are receptive to the sexual advances of males. This seems to be true for a number of non-primate mammals, and for the prosimians, tarsiers and some New World monkeys. However, Old World monkeys and apes have reproductive cycles which include both monthly menstrual periods and relatively long periods at other times of the year when they are not only receptive to advances of the males, but actively solicit sexual activity. These species do not fit the traditional concept of estrous cycles revolving solely around ovulation, since it is clear that estrus and ovulation do not coincide in many species. In the great apes, females come into estrus throughout the year, while in seasonal breeding monkeys, the females exhibit estrous behaviors several times during a specific season, and may even continue sexual activity after conception. Clearly, in these cases ovulation and estrus are not closely linked, and estrus is currently defined in terms of behavior patterns rather than physiological condition. Since it is virtually impossible to test hormonal levels and the timing of ovulation under field conditions, there is no way of establishing the point in the estrous cycle at which ovulation occurs or even if it does occur. Estrus in nonhuman primates is generally described as the female's motivation and willingness to mate, a definition that recognizes, not only the female's receptivity to the advances of the male, but also her **proceptivity**, referring to the female's active solicitation of sexual activity (Hrdy and Whitten, 1987).

126

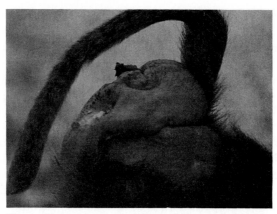

Figure 10.1 Extreme swelling of the perineal skin exhibited by an estrous female baboon. (Photo by Brian Keating. Courtesy of the Calgary Zoological Society.)

There is a great deal of interspecific variation in the external signs of estrous in female primates. Female baboons and chimpanzees show extreme physical changes in the form of reddening and extreme swelling of the **perineal skin**[1] (see Figure 10.1), while in Japanese macaques, the facial and perineal skin only become reddened. In species such as vervet monkeys and gorillas, who show no visible signs of estrus, behavioral signs are much more reliable indicators of the estrous state of the female. Estrous behaviors can range from distinctive vocalizations, irritability and heightened levels of activity, to very obvious proceptive actions such as following, grooming, genital presenting and even mounting a male. It has been noted that females who show the most extreme swelling of the perineal skin, such as baboons, chimpanzees and some colobine species, generally live in large multi-male, multi-female groups. Presumably, this is to advertise their estrous state, thereby increasing the mating opportunities and the possibility of having an infant sired by a "fit" male. Such striking indications of estrus may not be as necessary in monogamous and polygynous societies where there is only one breeding male in the social group. There is evidence in some primate species that the males tend to concentrate their sexual advances to females at mid-cycle. This is particularly evident in species in which the females show prominent sexual swellings, where the males attempt to copulate with females at the height of their **tumescence** (Hrdy and Whitten, 1987).

Copulation

There is ample evidence from the deprivation studies carried out by Harlow (see Chapter 13) that monkeys raised in isolation from their mothers and peers are unable to mate successfully. Mating behaviors must be learned in a social setting, and although the act of mounting is instinctive, the appropriate motor and social skills to complete a successful copulation can only be gained by observation and practice. Anyone who has watched young primates cannot help but notice that practice-mounting forms an integral part of their play patterns (see Figure 10.2).

In most primates, a male only needs to mount the female once for a successful copulation to occur. However, some primate species are referred to as "series-mounters" because the male must mount the female several consecutive times during a 20 minute period before ejaculation occurs (see Figure 10.3). This type of copulatory behavior is interesting in a social sense because it can be disrupted easily and requires sustained cooperation between the male and the female to be successful. It is not surprising that the males and females exhibit elaborate courtship rituals and tend to form consort bonds in species such as Japanese macaques, who are series mounters (Michael and Zumpe, 1971; Fedigan, 1992). A consort bond is an intense temporary bond between a male and a female based on sexual attraction, which is characterized by a number of preliminary courtship behaviors. The consort pair spend most of their time together, traveling, foraging, grooming, supporting each other in agonistic encounters and developing a rapport that contributes to the success of the mating sequence. These relationships can last for days or even weeks, and a female may consort with several males during a breeding season (Fedigan, 1992).

Figure 10.2   Juvenile Japanese macaques play mounting.

Figure 10.3   Adult Japanese macaques copulating.

Given that primates living in multi-male, multi-female groups mate with a number of members of the opposite sex, they are often said to be promiscuous in their sexual activities. However, since mating often takes place in the context of social bonding, as in the case of Japanese macaque consort relationships or baboon friendships, the term promiscuity, which suggests indiscriminate or random sex, does not really reflect the primate pattern of sexual activity. Soliciting behaviors, such as genital presents, displays and mutual grooming have been recorded in the males and females of almost every primate species, indicating that both sexes are choosing specific mates. Recent studies of female sexual activity have shown that females not only solicit males, but can avoid mating by walking away or simply sitting down (Smuts, 1987). Orangutans are the only species in which males have been observed forcing their sexual attention on anestrous females. Subadult male orangutans, who are much larger than the females, appear to be able to copulate with unwilling partners. However, these alliances never appear to result in a pregnancy (Galdikas, 1981).

Male Dominance and Mating Success

The assumption that males compete for a place in the dominance hierarchy, and that the access to estrous females is directly related to their rank, has persisted since the early days of primatology. The premise that dominant males are more active sexually assumes that they will be more reproductively successful and father the most infants. The relationship between male dominance and mating success has attracted a great deal of discussion and research, however the results of these studies have been extremely inconclusive. It is possible to find conflicting data regarding the mating success of dominant males from studies conducted on the same species, and in some cases, on the same study group. Although dominant males clearly have the ability to intimidate others, there are at least three factors that influence their access to estrous females: male-male coalitions, alternative strategies employed by subordinate males, and female choice of mates. Research on savannah baboons, in particular, has indicated that older, lower ranking males frequently form coalitions in order to compete with dominant males for access to females. Subordinate males often avoid open competition with dominant animals and resort to other strategies to secure copulations. Field studies have reported instances of low-ranking males in multi-male groups sneaking brief, opportunistic copulations away from the center of activity. In polygynous societies, solitary males simply follow a group, waiting for an opportunity to challenge and displace the resident male. The recent interest in female sexual behavior has indicated that female choice plays a major role in primate mating activity. Estrous females exert their choice of mates by actively soliciting copulations or by refusing to cooperate with the sexual advances of the male, and in some species the females are the main initiators of sexual activity (Smuts, 1987). The relationship between male dominance and reproductive success found in many mammalian species does not necessarily hold true for primates. Perhaps the behavioral flexibility, for which primates are noted, has allowed lower-ranking males and females to nullify the reproductive edge of the dominant male.

Homosexual behavior has been noted in some primates, in particular the females of three macaque species and female pygmy chimpanzees. Due to the sensitive nature of the subject, these behaviors were either ignored by early primatologists or reported as aberrant reactions to the stressful conditions of captivity. However, primatologists studying free-ranging Japanese and other macaques observing the frequency of sexual activity between females have concluded that it is a part of their usual behavioral repertoire. Fedigan (1992) reported that the females who showed the most homosexual activity in the Japanese macaques of the Arashiyama West troop were those that had lost their infants from the previous year, and were sexually active with both males and females. None of the females observed were exclusively homosexual. She suggested that the homosexual behavior in female Japanese macaques could function to promote the formation of year-round bonds with unrelated females as a potential source of support. Kano (1980) reported that female pygmy chimpanzees often embrace each other ventro-ventrally, and suggested that the frequent genital-rubbing behaviors of the females serves as a mechanism for relieving tension since it usually occurred in the context of competition for food (Kano, 1980). The homosexual activity observed in primates and other mammalian species can be viewed as one of their many behavior patterns that serve a function in group life. It is clear that sexual behaviors are used by both male and female primates for reasons other than reproduction. The act of mounting a dominant male by a subordinate animal appears to divert aggression, while other ritual mountings can be signs of excitement, reconciliation, appeasement, self assertion or merely friendship (Jolly, 1982).

Summary

Nonhuman primates exhibit an enormous amount of variability in the physical and behavioral manifestations of sexual activity. Some species mate throughout the year while many others have a discrete breeding season, during which all sexual activity takes place. The physical indicators of estrus, the term used to describe the period when the females are motivated and/or willing to mate, differ greatly between species. some females exhibiting reddened, swollen sexual skin and others showing no signs at all. Research has shown that both the males and females play an active role in sexual behavior in almost all nonhuman primate species and are discriminating in their choice of mates.

---

[1]Perineal skin is the area of skin between the ischial callosites in the females of some primate species, which encompasses the openings of the anus, the vagina and the urethra.

Chapter 11

Socialization

Introduction

Socialization is a concept that deals with the way in which the early social experience of an individual shapes future social behavior and acts to assimilate the young into group life. Frank Poirier (1972:8), who edited a volume devoted to primate socialization, described it as "the process that links the ongoing society to the new individual." and suggested that "Through socialization a group passes on its social traditions and ways of life to succeeding generations." **Process** is the key word in Poirier's definition because socialization is continuous and ongoing, and can occur at any phase in an individual's life. The term is usually applied to the social development of the young, but socialization can take place in adulthood as well. The function of the socialization process, from the standpoint of the society, is to mold the young into participating group members. However, socialization serves two important functions in the social development of the individual: 1. The young learn how to form and maintain affiliative bonds with group members; and 2. They are able to establish a behavioral repertoire of appropriate responses to a variety of social situations. There are a number of group members, such as parents, caretakers, kin and peers, who facilitate the socialization process. Although these individuals promote the socialization of the young in an efficient and predictable way, each individual enriches the society with its own unique characteristics (Fragaszy and Mitchell, 1974).

Since all primates are social animals who live in some type of organized social group, they require socialization for survival. Since newborn primates are relatively **altricial**, or helpless in terms of motor ability, they require unusually long periods of parental care compared to mammals of similar size. In many species of monkeys, the young depend on their mothers for almost a year, while the offspring of great apes are not weaned until they are about four years of age. Precocial infants are independent at an early age, as in the case of newborn ungulates who are able to get up and follow their mothers shortly after birth. The term "altricial," when used to describe primate young, is a relative one, since prosimian infants develop quickly and are far more precocial than young monkeys and apes.

Socializing Agents

Mother

The mother-infant bond is the first bond ever formed, and is generally recognized to be the most important bond in nonhuman primate society because it forms a pattern for all of the infant's future social relationships. The mother is clearly the most critical socializing agent for the very young primate. Primate females usually bear a single infant who is dependent on her for nourishment, transportation, comfort and protection. The socialization process begins with the reflexive stage in which the infant responds to stimulation in an indiscriminate way, clinging tightly to its mother's ventrum. The very act of clinging ensures the maintenance of contact with the mother and guarantees the formation of a deep maternal bond (Fragaszy and Mitchell, 1974) (see Figure 11.1).

Figure 11.1   Japanese macaque infant with its mother. (Courtesy Karen Dickey.)

Given the critical quality of the mother-infant bond, one would expect maternal care to be relatively consistent over most primate species. However, the extensive research that has been conducted on the mother-infant relationship in primates has shown that there is a marked variability in the nature and degree of maternal care both between and within species. The following factors may contribute to the variability in maternal care (Nicolson, 1987):

1. Social Learning

It was once assumed that the "maternal instinct" would surface as soon as a female bore an infant and that she would be a good and caring mother. Harlow's (1965) laboratory experiments with motherless female rhesus macaques proved graphically that female monkeys raised in social isolation were totally inadequate mothers. They appeared to have no conception of motherhood and either abused their new infant or ignored it completely. Since appropriate mothering is usually learned from one's own mother, the style of maternal care is often repeated in the daughter. Females who have had good mothers will tend to be good mothers themselves, whereas an abusive style of mothering will be likely be perpetuated in the daughter.

2. Temperament

Females with placid temperaments will tend to be more relaxed mothers than nervous and high strung females. These differences in temperament could be directly associated with some of the environmental or social factors that affect maternal care (e.g., dominance rank).

3. Social Rank

In species with female dominance hierarchies, the fact that high-ranking females tend to be more confidant than subordinate females and less subject to harassment could result in different styles of mothering. Although the evidence is not overwhelming, some field studies have shown that dominant females are more relaxed and less restrictive mothers than lower-ranking females (Berman, 1984).

4. Maternal Experience

Maternal experience, or **parity**, which refers to the number of infants a female has produced, has been cited as another factor which could affect maternal care, at least in the early months after birth. **Primiparous** mothers, who have only had the one infant, have been found to be more anxious and protective of their infants than **multiparous** females, who have had more than one offspring. Although the long-term effects of first-time mothering have not been established, there is some evidence that incompetent mothering by primiparous females could contribute to a higher mortality rate in their infants compared to those born of more experienced females (Nicolson, 1987).

5. Species Differences

Physical factors such as the precocity of the infant, the rate of maturation, birth spacing and gestation period have a significant effect on the duration of maternal care. Infanthood, which is usually defined as the period of infant dependency, varies from about six months in lemur species, to one year in monkeys, to three to five years in the great apes (Napier and Napier, 1985). Although the quality of maternal care may be similar, chimpanzee mothers will clearly spend much more time with their infants than lemur mothers.

6. Social and Physical Environment

The group social structure, which dictates the availability of peers and adults with whom to interact, will also affect the degree of maternal care required. In multi-male and polygynous groups, where a number of young are growing up together, infants will form play groups and spend less time with their mothers than in monogamous groups where siblings are the only playmates. Environmental adaptations can also contribute to the interspecific variability in maternal care. In arboreal species, infants must remain in contact with their mothers until they are physically able to locomote safely in the trees, whereas semi-terrestrial infants are able to locomote on their own and move away from their mothers at a much earlier age. A natural environment offers much more in the way of social and physical stimulation for the developing infant than a zoo enclosure. Restrictions in terms of space and the availability of playmates keep infants in zoos and laboratory colonies close to their mothers. In the wild, infants spend much of their time in locomotor activities or interacting with their peers, returning to their mothers only for food or comfort.

7. The Sex of the Infant

Some researchers who have studied the mother-infant relationship in captive macaque species have reported that mothers treat their male infants differently than their female infants. Females tend to spend more time away from their male infants, reject them earlier and wean them more severely than their daughters (Fedigan, 1992). Although these differences have been viewed as an adaptation to promote greater independence in male infants, two proximate explanations have been offered for the differential handling of male and female offspring. The mother could be responding to cues related to physical differences in the infants in terms of smell or external genitalia, or she could be reacting to inherent behavioral differences in the infants. It is possible that young males are rejected earlier and more severely because they are more active and difficult to handle than young females, due to the presence of male hormones. Although research on the behavior of captive rhesus macaque and baboon infants has indicated that the male does play more vigorously than the young females, hormonal studies revealed that neo-natal males have very low levels of **testosterone** (male hormone) (Goy, 1968). Since castrated and ovarectomized rhesus infants showed the same differences in activity levels as normals infants, the behavioral variation in young males and females is not based on hormones (Goy, 1966). There is some evidence that levels of male hormone in the developing fetus may affect not only the physical characteristics but also behavioral attributes in the young. Perhaps the level of prenatal **androgens** sets the limits for the expression of certain behaviors (see Fedigan, 1992, pp 165–168 for a discussion of the research into the pre-natal androgen levels). There is no conclusive evidence to suggest that the variability in maternal behaviors based on the gender of the offspring is due to differences in the mother's responses to her infants or to differences in the behaviors of the infants. In all likelihood, a combination of factors is involved.

As the infant develops, the attachment to the mother is not sufficient to satisfy the need for social stimulation, and the infant is ready for exposure to a non-maternal environment. The contribution of other members of the social group now becomes important in the socialization of the young primates.

Female Caretakers

Female caretaking, where a female other than the mother cares for an infant, was once referred to as "aunting." However, since this expression implies a biological relationship, which may or may not be present, the term "allomothering" is now widely used. Allomaternal behaviors can take many forms, ranging from grooming and muzzling, to carrying, playing, and even **allonursing** in some species (see Figure 11.2). In a number of primate species, such as the common langur, bonnet macaques, patas monkeys and

Figure 11.2 Japanese macaque female nursing her own infant and allomothering infants of other females. (Courtesy Linda Fedigan.)

vervets, allomothering is common occurrence, even in the first days after birth, whereas in chimpanzees, females other than siblings rarely care for another's infant. Allomothering is generally confined to a specific class of females within a primate society. In some species, the young **nulliparous** females (those who have not had an infant) do much of the caretaking, while in others, like ring-tailed lemurs, females with infants of their own will babysit (Nicolson, 1987).

A number of suggestions have been offered as to why a female would look after an infant that is not her own, particularly an unrelated infant. The primary explanation is that allomothering provides practice for mothering, allowing young nulliparous females an opportunity to learn maternal skills. At first, naive females often have difficulty dealing with uncooperative infants, but by the time they have had babies of their own, they will likely be competent caregivers. It has also been suggested that low-ranking females could benefit by caretaking infants of dominant females, thereby gaining greater access to resources and a reduction in the aggression directed towards them. Allomothering also benefits the the mother by allowing her some relief from the energy-expensive business of raising an infant. The female caretakers provide a "babysitting service," so that the mother is free to forage, feed, or even rest on her own. The infant, as the recipient of the caretaking, has an opportunity to widen its social horizons and form bonds with other females. This improves the infant's chances of adoption, should anything happen to its mother (Nicolson, 1987).

Clearly, allomothering can provide benefits to all the parties involved. The young females gain mothering experience, the mothers are given respite from mothering and the infants gain social bonds. There is, however, a possibility that the infant might suffer from incompetent handling by a nulliparous female or by deliberate abuse, and may even be kidnapped and die from malnutrition. Although cases of infant abuse and mistreatment by other females have been recorded, it has been my experience that primate mothers generally know the whereabouts of their infants and would do their utmost to intercede in an abusive situation.

An interesting study by Kaufman and Rosenblum (1969) illustrated striking species differences in allomothering. They raised two closely-related species of macaques (bonnet and pigtail macaques) in identical laboratory environments and observed the interactions between females with infants and other group members. The bonnet macaques were extremely sociable, clustering in small groups, and the mother were extremely **permissive** with their infants, allowing both males and females to caretake them. The pigtail macaques were much more solitary, and the females would not allow other group members near their infants. When the mothers were removed from the social groups for short periods of time, the exclusivity on the part of the pigtail females adversely affected their infants, because the other group members avoided them, leaving them lonely and depressed. In contrast, the infant bonnet macaques were unaffected by the separation from their mothers because other females immediately began to care for them. From these observations, the researchers concluded that the **exclusivity** of the mothers with their infants could be an important factor determining the degree of allomothering in a species. Other females, who would normally be interested in the infants of others, are prevented from allomothering because of the the restrictive behavior of the mothers.

Male Caretakers

Although male caretaking had been recognized in some primate species for a long time, very little attention was given to interactions between infants and adult males until relatively recently. Male care was ignored by researchers because it appeared to be uncommon and of no special significance to the survival of the infants. But today male caretaking of infants is regarded as a subject worthy of investigation, largely due to the recent interest in sociobiological theories and the adaptive significance of social behaviors (Whitten, 1987).

The wide range of male-infant interactions in nonhuman primates was evident from the the studies of early primatologists. Adult males in some species, particularly marmosets and tamarins, expend a great deal of time and energy in the care and raising of their offspring, while other primate males protect the infants and females from predators but rarely form bonds with the young. In other primate societies, the males are very tolerant of infants and often form close bonds with the offspring of particular females. The males of some species are intolerant of infants and discourage their advances. However, the most extreme case of infant abuse is the infanticide observed in the langurs of south India where the males in all-male groups attack the heterosexual groups and kill the infants.

The primary explanation for the expression of male infant care relates to the **parental certainty** theory developed by Trivers (1972), which argues that a male will expend energy in raising the young if he is sure that he is the father. Although there is evidence of paternal care in monogamous species, such as marmosets, tamarins and titi monkeys, where there is paternal certainty, the data supporting this theory in polygynous and multi-male species is less convincing. Although there is a high probability that the resident male is the father of the infants in species living in polygynous groups, little male care has been recorded. In multi-male societies, where there is the least proof of paternity, there are some striking examples of male caretaking. In Barbary macaques (*Macaca sylvanus*), where estrous females mate with several males daily, adult males give substantial care and attention to the infants from birth. Common baboon males, who live in very large multi-male groups, protect and help care for the infants of particular females with whom they have "friendly" relations (Strum, 1985). It would seem that the predictions of the parental investment theory do not fit the patterns of male care in many nonhuman primate societies. This is possibly because most male caretaking activities do not carry great costs in terms of reproductive success, nor large benefits to the survival of the infant. In the case of the baboons males, the care given to specific infants may actually enhance their ability to mate with their mothers (Whitten, 1987).

Mitchell et. al. (1974) tested the theory that male care is related to the exclusivity of the females by allowing captive rhesus macaques access to infants. They found that the male rhesus monkeys, who are normally disinterested in the young, virtually adopted the infants, playing with them, grooming them and protecting them. Although the isolation of a laboratory setting could have influenced the male rhesus monkeys' need for companionship, the results do suggest that male participation in the care of the infant may be due in part to the restrictiveness of the mothers.

Kin

A species must be studied on a long-term basis to fully understand the socializing effect of kin. The three species for which we have long-term data covering several generations are the rhesus macaques at Cayo Santiago, the chimpanzees at Gombe Stream Reserve and several troops of Japanese macaques. The rhesus and Japanese macaques have similar kinship systems articulated around the females. In these and many other multi-male, multi-female societies, the females remain in their natal groups for life and matrilineages of animals descended from a common female ancestor are built up. The males in these groups tend to peripheralize at puberty and often leave to lead a solitary life or to join a new troop. During the infant and juvenile stages, the young males and females learn to adjust to group life in close proximity to members of their matrilineage. Members of these kinship units treat each other in special ways. They groom each other more than outsiders and readily form alliances to defend each other in agonistic encounters. The young primates of both sexes grow up in a warm social atmosphere surrounded by their close kin, and the females form bonds with each other that last throughout life (see Figure 11.3). As yet, there is no accurate way to determine paternity in large multi-male groups where the females mate with a number of males, so it is difficult to assess the socializing effect of relationships through the males.

In chimpanzee society, the males remain in their natal group, and the females tend to transfer to new communities during their first or second estrous cycle. The basic kinship

Figure 11.3   A kinship group of Japanese macaques. (Courtesy Linda Fedigan.)

group is the matrifocal unit, made up of a mother and her subadult offspring. However, the close affiliation between related females found in macaque societies does not exist. Although groups of bonded males, some of whom are known to be brothers, travel together and support each other, young chimpanzees are socialized by members of their immediate family members rather than large kinship groups (Whitten, 1987).

As long-term data becomes available on more primate species such as baboons and vervets, it is becoming increasingly clear that kin are of major importance in the socialization of the young, particularly those living in large social groups (Fedigan, 1992).

Peers

Peer group interaction, which takes place in play, is a valuable socializing agent in primate societies when a number of infants are growing up together. Infants are attracted to each other, and in species like Japanese macaques, where the young are born during the same season, they form play groups as soon as they are mobile (see Figure 11.4). In the early years of primatology, play was ignored as a research subject, not only because it was thought to be a frivolous activity that served no important function, but because it was so difficult to define and record. However, despite problems encountered in data collection, play is one of the easiest of all animal behaviors to recognize. Even the most naive observer is able to identify play, primarily due to a **metacommunication** associated with playful actions. Metacommunication refers to a communication, operating at a subconscious level, that accompanies a communicative act and functions to identify the motivation behind a behavior. Thus, the open-mouthed play faces that accompany a bout of vigorous wrestling between juveniles, tells the observer that the wrestling is only in fun (see Figure 11.5). Play behaviors are repetitive, exaggerated, fragmented and often mimic adult activities such as mounting and fighting. There are some play behaviors, such as chasing, wrestling and fiddling with objects, that are common to all primate species, and some that are species-specific. A rapid sideways leap off trees that mimics the display of the adult male, is only seen in young patas monkeys, while the stiff-legged, pogo-stick style of hopping is typical of juvenile vervets (Fedigan, 1992).

Figure 11.4   Two young Japanese macaques at play. (Courtesy Karen Dickey.)

Figure 11.5   Young Japanese macaque confronting a peer with a play face. (Courtesy Linda Fedigan.)

Some primatologists feel that play does not represent a major factor in the socialization process because the young seem to develop into functioning adults even when play is rare or absent, as in times of stress when resources are scarce and play activities tend to cease. However, others view play as an important youthful activity and cite three possible functions of it:

1. Physical Development

Primate social play is highly active and includes all of the locomotor patterns observed in adults. The infants become more coordinated and competent at their motor skills through daily solitary or social play.

2. Practice for Adulthood

Primate play often includes rudimentary versions of mounting, mothering and fighting, all of which are essential activities in adulthood. Research has illustrated the importance of learning these skills in the presence of peers, at a time when the consequences of naivty are not critical to survival.

3. The Development of Social Perception

In many species of monkeys, infants are born with a distinctive **neo-natal coat**, which lasts from birth to about 10 weeks of age. During this period, adults are extremely tolerant of the infants' clumsy advances and inappropriate social behavior. However, during play young primates learn the consequences of their actions with playmates who also have neo-natal coats. These early encounters with peers serve to develop the skills of social communication that are so important in adjusting to life in a social group (Walters, 1987).

Although there is no consensus of opinion as to the adaptive function of play, it appears to be a useful socializing agent in species where peer group interactions are common.

The Mother-Infant Relationship

During the neo-natal period, it is important for the infant to have a close and satisfying relationship with its mother in order to be able to form new relationships easily (Fragaszy and Mitchell, 1974). The contact between the mother and her infant is intimate and intense in the early months. However, as the infant matures physically and socially, the mother-infant bond gradually weakens and an increasing amount of time is spent interacting with peers. By tracing the development of infant rhesus monkeys from birth to independence, Harlow and others (1963) observed three stages of maternal behavior. The first stage is characterized by maternal attachment and protection, during which the infant either clings to the mother's ventrum or is cradled in her lap (see Figure 11.6). The infant is carried on the mother's ventrum and is restrained if it tries to leave her side. In the next stage, the mother's ambivalence towards her infant is marked by alternating rejection and protection. Finally, the mother begins seriously rejecting her infant until it is fully

Figure 11.6 Common baboon female nursing her infant. (Photo by Brian Keating. Courtesy of the Calgary Zoological Society.)

independent. This phase often takes a long time, and in species where the females remain in their natal group throughout life, the separation from the mother is never complete. Clearly, these are not discrete stages with a definite beginning and end point, but represent a considerable overlap in maternal behavior. Although the pattern of the gradual relaxation of maternal care is applicable to most primates, the wide variation in the onset, duration and intensity of the stages is not surprising, given species differences in physiological, ecological and social characteristics.

Weaning the infant from the dependance on its mother, which begins at the ambivalent stage described above, appears to be a very traumatic time for young primates, as they often react to their mother's rejections with vocalizations and tantrums. Research has shown that the infants, who are developing physically and socially, are responsible for controlling the amount of time spent close to their mothers. Whether the initiative on the part of the infant is caused by the maternal rejections or is the result of their increased physical and social mobility has been the source of much debate. Both the mother and the infant are contributing to the achievement of independence to some degree, but the relative contribution of each has never been established (Collinge, 1991). Given that the the process of weaning a young primate from the nutritional and emotional support of its mother covers a relatively long period, the role of each individual changes, depending on the circumstance (McKenna, 1979). During mating season, when the rate of maternal rejections increases significantly, the mother is likely the major instigator of independence. However, when a play group of juveniles appears, the impetus reverts to the young. In any event, young primates do eventually achieve independence from their mothers, who, generally speaking, are very tolerant and caring of their young even during the weaning process.

Summary

Although the process of adapting the young primate to active participation in the social life of the group is accomplished by a number of individuals, such as kin, peers and caretakers, the mother is clearly the most critical socializing agent. The mother-infant bond, which is so strong in the early months of life in all primates, persists throughout life in the females of many female-philopatric species. Even though the mother-infant relationship changes over time, from intimate contact to relative independence, the quality of the bond can affect the future relationships of the developing primate.

Chapter 12

Social Behavior

Kinship

The importance of kinship in the social life of nonhuman primates was first established in Japanese monkeys by researchers at the Japanese Monkey Center and in the rhesus macaques at Cayo Santiago. Records of these primates date back to the 1950s and cover several generations of animals. Since the females in these macaque species mate with a number of males during a breeding season, paternity is unknown and genealogies are traced through the females. In these groups, where the females remain in their natal groups for life, animals who are linked by descent from a common ancestor are said to belong to the same matriline or matrilineage. Later long-term studies of wild chimpanzees provided additional information not available from these matrilineally-based societies. In chimpanzees, where the males remain in their home community for life and the females tend to transfer between groups, the main kinship group is the matrifocal unit, made up of the female and her offspring (Sade, 1991). It was originally thought that once an infant was weaned the mother-infant bond was completely severed. However, from the wealth of information on the social lives of Japanese macaques, rhesus macaques and chimpanzees, it is clear that the young of both sexes stay with their mothers until adolescence, and females in matrilineally-based species remain close to their mothers throughout life (Fedigan, 1992).

It has been shown that kinship has definite behavioral implications, as the members of matrilines act differently towards each other than to unrelated individuals. They tend to spend much of their time in close proximity to each other, grooming each other, traveling, foraging and sleeping together, and readily form alliances to defend each other in conflict situations (see Figure 12.1). The **alloparenting** of infants is more common between related animals, and members of kinship units groom each other more frequently than they do outsiders. Since grooming represents an invasion of an animals' personal space, grooming bouts with unrelated animals are often preceded by lipsmacks, presents for groom and other friendly overtures. The frequent bouts of agonism that reportedly occur between kin are usually at the level of bickering rather than all-out fighting. The benefits of kinship are usually discussed in terms of the females, however in chimpanzee society, brothers often travel and socialize together, and support each other in confrontations with dominant animals (Gouzoules and Gouzoules, 1987). Goodall (1986) suspects that the members of the all-male groups of chimpanzees, who are clearly bonded, may be related in some way.

Figure 12.1   A kin group of Japanese macaques.
(Courtesy Linda Fedigan.)

In Chapter 11, kinship was described as an important socializing agent for young primates, particularly those living in large multi-male groups. The developing infants interact regularly with relatives, other than their mothers, and grow up in a warm social climate with their brothers, sisters, aunts and grandmothers. They have ample opportunity to learn the social skills necessary for group living, and Gouzoules (1980) has suggested that infants with more kin become independent from their mothers at a younger age. Although primates growing up in monogamous societies do not have the extensive kinship networks of those living in large groups, young marmoset and tamarins interact closely with their younger siblings, carrying them and sharing food with them.

Some large multi-male societies, such as Japanese and rhesus macaques, have a tendency to split into two groups when they reach a certain size. Although one might expect such a split to be characterized by a random break-up of the social group, these group fissions often occur along kinship lines, with whole matrilines staying together and moving to a new territory. Consequently, the members of each group are more closely related to each other than to their conspecifics in the other group (Fedigan, 1992).

Kinship and Dominance

Thanks to the diligence of Japanese primatologists, who were particularly interested in the social networks of their indigenous monkeys, we are able to understand how kinship affects dominance rank. In Japanese macaques and other large multi-male societies, offspring rank according to their mother's rank, and infants of high-ranking females will also be high-ranking. This rule refers particularly to females who remain in their social groups for life, since the males, who tend to move away, may have to reestablish their rank in a new group. Siblings rank in reverse order to their age, perhaps because the mother tends to defend her youngest offspring over the others, and the mother almost always ranks above her daughters. When the mother dies, the oldest daughter takes over her position as the head of the family. It has also been established, at least in Japanese and rhesus macaque societies, that whole matrilines achieve dominance over others, with the kinship group of the alpha female being the highest-ranking family. Generally, dominance hierarchies between matrilines are fairly stable, however Gouzoules (1980) documented a rank change in a kinship group in the Arashiyama troop of Japaneses monkeys. A female from a low ranking lineage challenged the ailing alpha female for her position. With the help of her relatives and the alpha male, the challenger won the power struggle, became the alpha female and, subsequently, her entire matriline rose in rank. Although the alpha female has since died, her daughter has taken over her position and the kinship group remains dominant.

Kinship and Breeding Patterns

The literature on nonhuman primate mating practices records inbreeding avoidance in most species, particularly in the case of mother and son. Fedigan (1982) reported, that in over 1000 observed matings in the Arashiyama group of Japanese monkeys, none were between mother and son, and only one was between brother and sister. A number of studies of non-primate species have shown that copulations between relations has resulted in "inbreeding depression" (Gray, 1985:196). The offspring of liasons between close relatives often had a lower reproductive potential because of smaller size, birth defects or reduced fertility. The few studies that have addressed this issue in nonhuman primates have also found evidence of reproductive depression (O'Rourke, 1979; Packer, 1979; Ralls and Ballou, 1982).

If reproductive depression results from inbreeding, then evolutionary theory would predict that animals would have been selected to avoid mating with kin in order to improve their reproductive success. Research has suggested that there may have been selection against inbreeding. Mating with kin occurs much less frequently than would be predicted for random mating (Gray, 1985). There are four proximate mechanisms observed in primates that could act to prevent inbreeding (Gray, 1985):

1. Dispersal patterns - Inbreeding is prevented in many species living in multi-male groups by the tendency for males to transfer between groups. However, it is interesting to note that even in semi-free-ranging groups, like the Arashiyama troop of Japanese monkeys in Texas, where males are forced to remain in their natal group, there appears to be a clear avoidance of mating with related animals.

2. Dominance relations between mother and son - Sade (1968), who studied mating patterns in rhesus macaques, suggested that males are inhibited from mating with their mothers because they remember the early mother-son relationship when their mothers were dominant over them.

3. The familiarity hypothesis - Others have argued that the aversion to mating with close relatives found in primates arises in response to the prolonged period of maturation and socialization which is spent in close proximity to their kin. This is an extension of the theory put forth by Westermark in 1898 to explain

the incest taboo in human cultures. He suggested that human incest prohibitions are not merely cultural artifacts, but are based on an aversion of mating with peers who have been raised together.

4. Female choice of mates - The theory that females often choose to mate with unfamiliar males to avoid inbreeding has some support in the literature. Although female baboons have been shown to follow the pattern of choosing males that have transferred from another group (Packer, 1979), inbreeding avoidance is not universally believed to to be the causal agency (Gray, 1985).

Although the evidence from the field studies of a number of primate species clearly indicates a general avoidance of mating between closely-related individuals, there is no consensus as to which of the above-mentioned explanations is most plausible. Gray (1985:201) summed up the problems involved in reaching such a consensus: "Given the interspecific variation in inbreeding, it is unlikely that a unitary explanation linking mate selection, inbreeding avoidance, specific proximate mechanisms, and inbreeding depression can be formulated."

Summary

In several species of Old World monkeys and chimpanzees, kinship has proven to be an important factor in determining patterns of social behaviors, such as allomothering, grooming and defense in conflict situations. Since paternity is difficult to establish in large multi-male groups, kinship is determined by relationships through the females, and discussed in terms of matrilineages or matrifocal units. In species in which the males transfer between social groups, the role of kinship in social patterns is more obvious in the females, who spend a large proportion of their time interacting with each other. In these societies, since the young grow up in close proximity to their close relatives, kinship groups represent important socializing agents. Given the fact that relevant data are unavailable, the role of kinship in the development of the young and the social patterns of species living in monogamous and polygynous groups may not be nearly as critical (Gouzoules and Gouzoules, 1987).

Aggression and Affiliation

The advent of World War II and subsequent hostilities has generated concerns about the instinctive nature of aggression in humankind and the subject has attracted the attention of both sociologists and animals behaviorists. A number of books on the subject, written after World War II, have given a misleading impression about the relationship between animal and human aggression. Lorenz (1966) wrote <u>On Aggression</u>, in which his main message was that humans and animals do have a killer instinct which is directed against members of their own species. Others have asked: "If aggression is a universal trait of social animals, then how can we explain the widespread tendency of animals in a fight to use restraint and not fight to the death, even when the most aggressive individuals control resources such as food or mates?" (Drickamer and Vessey, 1982:274). Living in a social group provides a number of benefits in terms of protection from predators and improved access to food resources, but also carries with it the problem of getting along with group members and coping with the competition for resources. Aggression is a conspicuous feature of life in primate societies. However, the use of restraint in aggressive encounters is equally obvious. The displays, threats and fights act to intimidate others, and may even cause injury, but rarely result in the death of the group member.

Prior to the 1960s, most primatologists would have said that nonhuman primates were extremely aggressive animals. This notion, fostered by anecdotes and folk tales, was confirmed by the scientific research of Dr. S. Zuckerman (1932). His report of the unbridled aggressiveness of the hamadryas baboons in the London Zoo was widely accepted as a general primate trait. The baboon group consisted of six females and a large number of males, who were strangers to each other. The males fought over access to the females, killing each other and the females in the process. And although more females were subsequently added to the group, the bloodbath continued until the few remaining females were removed. It was not realized that the chaos in the London Zoo was the result of a totally inappropriate social grouping until Kummer studied hamadryas baboons in the field. A normal heterosexual social group of hamadryas baboons consists of a single male and several females. Groups of closely-bonded males, who travel together, will challenge the resident male for his place in the heterosexual group, but are inhibited against disrupting a pair bond between one of their group members and a female. The severe disruption of the normal sex ratio in the hamadryas group in the London Zoo, and the placing of strange males together in the same enclosure, was certain to trigger unchecked aggression.

The idea that primates were aggressive animals was challenged when field studies in the 1960s and 1970s emphasized the peaceful nature of their study groups. Schaller (1965) stressed the gentleness of the gorillas in the Virunga mountains, while Jay (1965) and Goodall (1965) wrote about the amicable relationships found in the langurs of Northern India and the chimpanzees of Gombe. However, the illusion of peaceful primate societies had to be revised when primatologists began observing disturbing episodes of aggression and killing, primarily in the langurs of southern India and in Goodall's chimpanzees.

The first inkling of chimpanzee aggression arose as a result of a provisioning strategy, when bananas were put out for the chimpanzees at Goodall's campsite to entice the animals away from the trees so that they could be observed and identified. Although the strategy served its purpose as far as the research was concerned, the resulting aggression between the chimpanzees and against baboons who had also been attracted to the site, forced them to abandon the idea. In the 1970s, further aggression was reported by Goodall when the males in a large community of chimpanzees systematically attacked and killed all of the members of a small splinter group and took over their territory. Although these killings and the bizarre murder and cannibalism of infant chimpanzees by a female and her daughter ceased, the notion of the tranquil tenor of life in chimpanzee society was effectively dispelled.

The reports of male infanticide in the langurs at three sites in southern India also acted to shatter the concept of the harmonious quality of life in other primate groups. All-male groups of langurs were observed attacking the uni-male groups, ousting the resident male and competing for access to the females, whose infants had been presumably killed by the attacking males. Although the aggressive behavior of the South Indian langurs has been well documented, the causal agency for the killings is still being discussed. Jay has blamed the heightened level of aggression on the overcrowding and habitat disturbance in the South Indian sites, while Hrdy (1977a) has argued that the infanticide is a male strategy to improve reproductive success.

Although our impressions of the aggressive nature of nonhuman primates have changed dramatically over the years, the extensive body of data on primate social behavior has indicated that there is a great deal of interspecific and individual variability in the levels and manifestation of aggressive behavior. The scope and complexity of the concept of aggression makes it a difficult subject to address. Aggressive actions range all the way from stares and threats to bites and mortal wounds (see Figure 12.2). Ethologists discuss aggression in terms of behaviors directed towards another in which the goal is injury or harm, while dictionaries refer to aggression as attack behaviors. While these definitions describe the offensive nature of aggressive behavior, the universal concept of the term is ambiguous.

Figure 12.2   A double threat typical of capuchin monkeys. (Courtesy Katherine C. MacKinnon.)

In the world of trade and commerce, aggressive or self-assertive behavior is often admired as the mark of a successful individual. Agonism is a term that can be used more precisely than aggression to describe conflict behavior in animals. Agonistic behavior, as defined by Scott (1966), is any conflict behavior, either self-defensive or self assertive, directed towards a member of the same species. Agonistic acts directed towards members of the same group are usually discussed as dominance interactions, while agonism towards other groups of conspecifics is referred to as territoriality. Scott specifically excluded predatory or interspecific aggression from his definition of agonism because it stems from different behavioral systems, such as hunger. Although the term "agonism," which does not infer motive or intent, is preferable to "aggression" in terms of animal behavior, the following section will use the terms interchangeably to indicate any kind of self-assertive behavior.

Causes of Agonistic Behavior

Agonistic behavior in primates can be observed in a number of different contexts and can stem from a variety of sources. The factors that contribute to animal aggression can be discussed at the **proximate, ontogenetic** or **ultimate** levels of causation.

1. Proximate explanations for agonism deal with immediate causes arising from stimuli from the physical and social environment. Although competition for resources or mates is generally seen as a driving force in animal aggression, much of the agonism involves the establishment and maintenance of dominance which indirectly relates to access to resources (Chapais, 1991) (see Figure 12.3). Any disturbance of the social norm will result in tension in the social group and high levels of agonistic behavior. One disruptive situation that is almost certain to trigger agonistic behavior is the introduction of a strange animal into the social group. The strongest reactions come from the same age and sex category as the intruder. Bernstein (1964), who studied aggression in rhesus monkeys, reported that the reaction of adult males to the introduction of a strange adult male was swift and intense. The confrontation, which began immediately with displays, threats and bites, was usually short-lived, dissipating when one of the animals retreated from the conflict. In contrast, females displayed low-level agonistic behaviors, such as threats, scratches and puncture wounds, towards strange females. These agonistic encounters often lasted for several hours until

Figure 12.3   An agonistic encounter in a Japanese macaque group. The female on the right screams and runs away from the threats of the male on the left. (Courtesy Karen Dickey.)

some sort of social order was established. The harassment of an infant by a conspecific is another proximate cause of aggression in primates. A frightened infant will trigger an immediate agonistic response from the mother and related animals towards the cause of the trouble.

2. Ontogenetic, or developmental, factors can also be the causal agencies for agonistic behavior. Adolescent males reaching puberty, who have high levels of male hormones circulating in their bodies, are noted for their aggressive temperaments and their tendency to fight at the least provocation. Animals who have had an aberrant upbringing by an abusive mother also tend to be violent and aggressive, particularly females who almost surely will repeat the abusive behaviors patterns with their own infants.

3. Ultimate explanations emphasize factors that influence the evolution of aggression in primates. The primary explanation for agonistic behavior at the evolutionary level relates directly to the principle of epigamic selection in the theory of sexual selection discussed in Chapter 9. Males compete with each other aggressively for access to estrous females in order to maximize their reproductive success. The primate example that illustrates this causal agency is the infanticide observed in the langurs of southern India, where the males in the all-male groups attack the heterosexual groups, kill the infants, expel the resident male and then compete with each other for copulations with his group of females. This example, which emphasizes competition for mates as a factor triggering aggressive behavior, illustrates that explanations of behavior at different levels of causation are not mutually exclusive, but depend on one's point of view.

Sex Differences in Agonistic Behavior

A superficial glance at a primate social group might lead one to believe that males are more aggressive than females, especially in species where the males are much larger and more conspicuous. Much of the literature on primate aggression concentrates on male-male competition, which also tends to promote the idea that primate males are more aggressive than females. However, when discussing sex differences in agonistic behavior, it is less misleading to look at differences in the context, motivation and expression of the agonistic behaviors of males and females than to make general comparisons of sex differences in the level of aggression. Bernstein's (1974) research on aggression in rhesus macaques discussed above shows that the expression of agonism towards strangers of like sex and age varies dramatically between males and females. Where the males react immediately with violent, but short-lived aggression, the females can harass an intruder to death with persistent scratches and puncture wounds. There are other instances of obvious sex differences in the display of agonism. Mothers and female kin show an immediate agonistic response to the harassment of an infant, whereas males often show no response. In addition to situations where the aggressive responses of males and females differ, there are species in which the females are reported to be more aggressive than the males in their actions against like-sex offspring and territorial displays (e.g. callitrichids, ring-tailed lemurs and gibbons) (Fedigan, 1992).

The Function of Aggression

Too much aggression is obviously dysfunctional, as in the case of the hamadryas baboons in the London Zoo, and inevitably leads to chaos and the eventual demise of a social group. However, the controlled aggression that plays such a prominent role in the daily lives of most species must have a positive function in nonhuman primate society. There are numerous reports of instances where agonistic behaviors have been carefully manipulated by individuals to reap social benefits. Self-assertive behavior can help the individual maintain his or her rank in the dominance hierarchy and promote access to food resources and mates. In terms of group benefits, one of the main functions of agonistic behavior is resolving conflicts and restoring social order. *Control interference* is the term used to describe a situation in which a dominant male acts to break up a fight between other adult males. Franz de Waal (1989) suggested that aggression is necessary in order to effect compromises in tense situations. Without the aggressive encounter, conflict between individuals would remain unresolved and tension in the group would continue to escalate. Aggression followed by a reconciliation of the participants in the conflict serves to strengthen bonds that may otherwise have been severed. Agonistic encounters between neighboring groups of primates functions to allocate resources by dispersing the animals in space.

Aggression and Reconciliation

Aggression in nonhuman primates has been extensively researched and is the subject for countless articles and books. However, de Waal (1989) argued that the reconciliation after an agonistic encounter is equally important. The benefits of aggression are counter-balanced by the costs of fighting, in terms of injuries and social disorder. Conciliatory behaviors have been observed in all primates species to provide checks on fighting which would disrupt the social group if allowed to continue. Inhibitions against all-out fighting become very noticeable in large, complex social groups where social control is critical.

Although reconciliation is common to all primates, the expression of conciliatory behaviors is extremely variable, depending on the species and the relationship between the individuals who have been in conflict. In Peacemaking Among Primates (1989), for example, de Waal has documented the widely different peacemaking strategies in three closely-related macaque species and two species of chimpanzee. Rank differences in antagonists often dictate the type of reconciliation behavior, especially in species like rhesus macaques, where the dominance hierarchy is rigidly enforced. In the large social groups found in macaques, ruptured kinship bonds must be quickly repaired, and often bouts of intense grooming follow fights between related animals.

Primates are particularly **tactile** animals, and the most common and direct conciliatory gestures are those that involve some sort of personal contact between the fighting parties. Grooming, in particular, appears to defuse agonism and is part of most efforts of reconciliation. Common chimpanzees exhibit a number of very human affiliative behaviors, such as touching, hugging and kissing in their efforts to appease an opponent. In bonobos, the sexual presenting, mounting and genital massaging seen immediately

151

after fighting is evidence of the use of sexual contact to defuse agonistic situations (de Waal, 1989).

Alliances

The outcome of aggressive encounters in a primate society is often affected by the interference of a third party. Alliances are formed when an outside individual acts to aid the defending or the attacking party. The fact that many species exhibit stereotyped solicitation behaviors which act to enlist the aid of the third individual, suggests that alliances are basic to primate society. In macaques, baboons and chimpanzees, antagonists can be observed directing glances at other animals, inviting them to help out in a conflict situation. The frequency of alliances depends on the species' social structure as well as factors relating to the age, reproductive condition and the relationship of the individuals involved (Walters and Seyfarth, 1987). There are also a number of different types of alliances, depending on the intended outcome. Resource-specific alliances can be seen in free-ranging savanna baboons, where older and lower-ranking males frequently form coalitions to displace dominant animals from copulations with estrous females. These alliances appear to be reciprocal, in that the partners take turns mating with the females (Packer, 1977; Smuts, 1985; Chapais, 1991). In multi-male societies, like Japanese and rhesus macaques, where the females build up large matrilines, kin regularly defend each other in agonistic encounters. In species where the males remain in their natal groups, like chimpanzees and red colobus monkeys, related males will join together to challenge a dominant animal or to compete for access to estrous females. Alliances between males and females occur mainly at mating season in the context of consort bonding (Walters and Seyfarth, 1987). Chapais (1991:202) refers to **xenophobic alliances,** in which all the members of a primate troop will band together to defend their territory against another group of conspecifics.

Both primates and non-primates use alliances for the same basic reasons: defense of kin, maintainance or acquisition of dominance, and to gain access to resources. However, primates appear to be unique in the ways that they use coalitions to obtain some of these benefits. Research has shown that primates reciprocate in their cooperative efforts and go out of their way to cultivate supportive relationships with dominant animals. These are behavior patterns rarely seen in non-primate species. The behavioral complexity of primates, compared to that of other **taxa,** is nowhere more evident than in the social strategies used to cement relationships that will be useful in present and future contests (Harcourt, 1992).

Summary

The term agonism, which refers to any conflict behavior directed towards conspecifics, is often used to describe aggressive behavior in animals because it is free from implications about the motivation behind the conflict. Agonistic behavior is widespread in many nonhuman primate species and clearly plays an important role in the their daily lives. Competition for food, space and mates, either directly or indirectly through the establishment of dominance, appears to be one of the driving forces behind primate aggression (Chapais, 1991). However, there are a number of other situations that can trigger agonistic behavior in primates: the introduction of a stranger into the group, an attack on an infant or a close relative, and the increased level of male hormone in adolescent males.

Although unrestrained aggression would be dysfunctional, carefully controllled agonism can be viewed as beneficial in the social lives of primates. Primates use agonism judiciously to establish or to maintain dominance, and often dominant animals maintain social control by breaking up conflicts between group members. Franz de Waal (1990) suggested that some measure of aggresssion is necessary to "clear the air" and resolve tense situations. He also argued that the universality of conciliatory behaviors in primate societies indicates that reconciliation is equally as important as agonism. Since the social cohesion so necessary to primate societies rests on the maintenance of bonds with kin and unrelated animals, the mending of bonds ruptured by conflict is critical.

Returning to questions about the instinctiveness of animal and human aggression raised at the beginning of this section, research has shown that primate aggresssion is not innate and inevitable. Primate aggression is generally context-specific, occurring in competetive situations, in disruptions of the social norm, or in defense of close kin. There is no evidence of the accumulation and subsequent release of aggressive energy, suggested by Lorenz. The evidence, instead, points to the use of aggression and reconciliation to strengthen the ties that bind primate society together (de Waal, 1990).

Dominance

The individuals who win the most agonistic confrontations in a variety of situations are viewed as dominant. The concept of dominance allows researchers insights into the "power politics" that seem to be an integral part of primate social life. Although the importance of dominance rank in the day-to-day lives of primates is not clear, a high-ranking position does provide some benefits in terms of access to resources. Early primatologists who studied Japanese macaques and baboons were struck by the confident carriage of the dominant males and concluded that dominance must be related to power. However, since peaceful coexistence is so important in large social groups, dominant animals must allow lower-ranking individuals opportunities for appeasement, so that the losers are able to live with the winners. de Waal (1987:422) stated that "By regarding dominance relationships as a compromise between inevitable agonistic tendencies and a need for cohesive social groups, one gets a feeling for the precarious equilibrium that exists between the two."

Most of the discussions of dominance begin with a reference to Schjelderup-Ebbe's (1922) classic work on the "peck-order" in domestic chickens, which showed that strange hens placed together would peck each other until a linear hierarchy was established. The dominant or alpha female was able to inflict more pecks than she received and the "peck-order" theory of dominance was defined as the ability to intimidate others in conflicts. Although a functional definition of dominance revolves around the priority of access to resources such as food, space and mates, most researchers refer to dominance as power over others established through intimidation. In primates, dominance hierarchies are usually linear with the alpha male or female at the top, followed by the beta animal and so on down the line. The term "dominance rank" refers to an individual's position in the linear hierarchy.

Measuring Dominance

There are a number of ways of measuring dominance in primates, each reflecting different views of concept. The following discussion includes a description of the various methods and their validity in terms of assessing dominance in primates (Fedigan, 1992).

1. Monitor Fights

The dominance hierarchy can be worked out by observing the social group over a period of time and recording the winners and losers of fights, much in the same way that Schjelderup-Ebbe established the peck order in chickens. The dominant animal is the one who is able to defeat all of the others. Although this method may be suitable for determining dominance in domestic hens and other species, many primate fights are concluded without establishing a clear winner, and any decision about the dominant animal may be fairly arbitrary.

2. Food Test

In this type of test, commonly used in laboratories to establish dominance, a competition for food is created between two individuals Although the shortcomings of this method are well known, the ease of administration compensates for its liabilities. The main difficulty lies with the **dyadic**[1] approach, in that a competition between only two individuals rarely occurs in a primate social group. In most social situations, especially in polygynous and multi-males groups, there are often at least two participants and several onlookers who may affect the outcome of any encounter. A number of factors that could influence the result of the food test are: unequal physical abilities of the subjects, food preferences, the state of hunger, or simply the proximity to the food item.

3. The Frequency of Agonistic Signals

Since the incidence of physical fights may be relatively rare in some primates species, threat behaviors are often used as indicators of agonism. In this test, the animal exhibiting the greatest frequency of threats is judged to be dominant over the others. This is a poor way of establishing dominance in primates because the individual giving the most threats will likely be a very low-ranking animal, who has been harassed and is insecure in the company of other animals.

4. The Direction of Agonistic Signals

Instead of using the frequency of agonistic signals, this method of measuring dominance relies on recording who is threatening whom and who submits to whom in an agonistic encounter. A matrix of the frequency of threat behaviors directed by the individuals involved will indicate clearly who is the dominant animal (see Table 12.1).

Animal B was able to threaten all of the others, was therefore identified as the dominant animal. Using this criteria, the dominance hierarchy was established as B,A,C,D, with E as the lowest-ranking animals because of its inability to threaten any of the others.

Table 12.1 Matrix of the frequency and direction of agonistic signals

	A	B	C	D	E
A	X	0	111	11	1111
B	111	X	1	1	1
C	0	0	X	0	1111
D	0	0	111	X	0
E	0	0	0	0	X

5. Supplantations

The assumption in this test of dominance is that animals can be ranked according to who will be move away and be supplanted by whom. In this case, no overt signals are exchanged and the only sign that one animal is intimidated by another is that it is displaced from its present space by the approach of the other animal. The subordinate animal avoids a conflict situation by allowing the dominant individual access to its space. This is a useful test for dominance in primate species, such as vervets, who show few agonistic signals. The assumption on the part of the researcher about the motivation for the retreat of the "subordinate" animal could be viewed as a possible drawback to this method of assessing dominance.

A problem common to all of these methods is that they ignore the role of coalitions and alliances in determining the outcome of a conflict. Two animals acting together may be able to dominate an animal that neither could overcome alone—a situation which occurs frequently in baboon society. Gouzoules (1980) recorded an instance in the Arashiyama West troop of Japanese monkeys where a low-ranking female was able to overcome the alpha female by soliciting the help of the alpha male. Kawai (1965), a Japanese primatologist who studied dominance in Japanese monkeys, distinguished between basic rank based on individual prowess and dependent rank, which depends on an association with another animal.

Misconceptions About Dominance in Primates

The following are a number of misconceptions about the concept of dominance in nonhuman primates and the part it plays in their social lives (Fedigan, 1992):

155

1. Dominance is based on size and strength

Although dominance is associated with the ability to intimidate others, high rank in primates is often is related to age and maturity and not necessarily based on size and strength as in other mammalian species. It is not unusual to read about a dominant female who is old and toothless, or an alpha male who is suffering from a terminal illness but still retains his position in the group. Although a healthy body is helpful in agonistic situations, the ability to dominate depends on social awareness and the capacity to mobilize support from others if necessary. Dominant animals often have to be socially adept to maintain their rank in the face of challenges from aspiring conspecifics.

2. Dominant animals are the most aggressive

Although dominant animals may be confident, self-assertive individuals who do not avoid conflicts, they do not have to resort to aggressive behaviors in order to maintain their rank. Lower-ranking animals, who are frequently harassed, often have to depend on threatening those below them in the hierarchy in order to achieve access to any resources. Contrary to the view of dominant animals as the aggressors, it is often their role to reconcile conflicts between lower-ranking animals.

3. Dominance ranks are permanent

Although shifts in rank may be slow to take effect, there is ample evidence from field studies that dominance hierarchies may be relatively fluid. Goodall (1989) has reported a number of changes in the male dominance hierarchy in the chimpanzees of Gombe since 1960 due to challenges from younger, lower-ranking individuals. In Japanese and rhesus macaques, where the females remain in their natal groups for life, female dominance hierarchies are relatively stable, and the dominance structure in males changes frequently due to their tendency to emigrate to new groups. In these societies, whole matrilines may change in rank due to a disruption in the individual female dominance hierarchy.

4. Dominance is a personal attribute

Since dominance can only take place in a social context, it is not a personal trait that remains with an individual throughout its life. Dominant males who emigrate and join another group are not likely to retain their rank and will have to work their way up the hierarchy in the new society.

5. Dominance can be inherited

The theory that dominant animals will win competitions for estrous females, sire the most infants and pass on the genetic basis for dominance is suspect for a number of reasons that were discussed in Chapter 10. The point to be made here is that, although a healthy, well-functioning body will be an asset in the achievement of high rank, dominance, per se, is an attribute that only functions in a social setting and is not an individual trait that can be genetically transmitted.

6. Dominance hierarchies are confined to males

The early notion that stable dominance hierarchies in primates are only found in males has been disproved in the light of subsequent field work. In large multi-male groups, it is common to find clear evidence of dominance hierarchies in both males and females, while in polygynous societies, the tenure of the resident male often rests with the dominant females. Long-term studies of Old World monkeys, particularly baboons and Japanese and rhesus macaques, have indicated that the dominance hierarchies in the females who remain in their home groups are much more stable than those of the males who tend to transfer between groups (Fedigan, 1992).

Summary

Since it was impossible to ignore the power of the large, confident males in the species studied by the early primatologists, it is not surprising that the concept of dominance in nonhuman primates has long been a favorite topic for research and discussion. Over the years, long-term studies of Old World monkeys and apes have shown that many of the first impressions exaggerated the significance of the conspicuous dominant males to their societies. Dominance hierarchies extend to both males and females and it is not necessarily the largest, most imposing animals who are dominant over the others. The factors, other than physical prowess that appear to be more important in the determination of dominance are social awareness and the ability to mobilize support from kin and other social networks within the group. Although research has often shown that rank is not necessarily an important variable in **fecundity,** mate choice and many other social phenomena, at the very least, confer benefits in the allocation of scarce resources.

To conclude the chapter on primate social behavior, it would seem appropriate to quote Franz de Waal (1989:83), who aptly summed up the give and take in primate social life: "These ingredients of primate social life—coalitions, conflict resolution, social tolerance and strategic thinking—seem tightly interwoven, each one stimulating the development of the rest."

---

[1]A dyad refers to two animals, therefore a dyadic interaction is one between two individuals. Triadic refers to interactions between three individuals.

Part IV

Primate Intelligence

Chapter 13

Learning

The Nature-Nurture Controversy

Two very distinct schools of thought concerning the nature of behavior have provoked a considerable amount of discussion over the decades. The controversy revolved around the nature-nurture debate over whether behavior is instinctive and genetically programmed or is the product of learning and experience. European ethologists, notably Karl Lorenz (1956), emphasized the instinctive quality of behavior. He described behavior as a sequence of events, beginning with a stimulus that triggered an innate mechanism. This, in turn, programmed the animal to react with a stereotyped, genetically-encoded response that was entirely independent of learning. On the other hand, psychologists studying animal behavior in the laboratory stressed the experiental aspects of behavior and argued that behavior is the result of an accumulation of learned acts. Skinner (1938), an influential American psychologist, was a major advocate of the view that animal behavior can only be studied in terms of repeated responses to the same stimuli. He believed that stimulus-response connections are learned by the animal if they are continually reinforced. Psychologists like Skinner, of the behaviorist tradition, disagreed with the idea that many behaviors are generally predetermined and unstoppable, while ethologists deplored the mechanistic view of the behaviorists, which implied that behavior can be predicted and resistant to control. Scientists studying behavior have since modified their positions, and no longer believe that behavior is completely learned or completely instinctive. It is now understood that behavior stems from the interaction between genetic propensities and internal and external environmental agencies.

The word **instinct,** which refers to biologically determined behaviors, has fallen into disfavor because it carries overtones of predestination and the idea that behaviors are invariable despite environmental factors. "Instinct" has largely been replaced in the literature by the term **innate,** which refers to the biological potential to behave in a certain way depending on environmental influences. Innate, unlike instinct, is a relative term because it allows room for variability in behavior. Although some behaviors are strongly innnate and others have a significant experiental component, most behaviors fall on a continuum between the two extremes. Innate behaviors, such as rooting and suckling, are universal throughout a species or taxonomic group, and are usually recognizable because they are stereotyped in motor patterns. In contrast, learned behaviors are not as easily identified because they are diverse in motor pattern and show a great deal of individual variation. A behavior under a high degree of genetic control is likely to be common to all members of an order, or even to like age and sex categories within a species, despite variations in the environment. The biological and environmental agencies influencing behavior are so closely interrelated that the relative contribution of each is difficult to assess (Fedigan, 1992).

Although the **genotype,** which is the genetic make-up of an individual, sets the limits for the body, factors such as disease, nutrition, learning and experience have an effect on the behavior patterns of the individual. The **phenotype** is the term used to describe the sum of the anatomical and behavioral characteristics of the individual produced by the interaction between the genotype and environmental agencies. Variations in behavior can only occur within the limits set by the genotype. Therefore, certain behaviors are harder for some species to learn than for others. For example, in the laboratory, rats can easily be taught to bar press with a paw and pigeons readily learn to peck at keys with their beaks because these actions are part of their natural behavior patterns. It would be as difficult to teach a pigeon to press the key with its foot as it would be to force a rat to bar press with its nose, since pigeons peck naturally and rats do not.

Social Learning

Dr. K.R.L. Hall, a British psychologist, was the first primatologist to recognize the importance of social learning in primates. Since all primates are social animals who are adapted to group living, all of their learning in the developmental period and throughout life takes place in a social context (Hall, 1968). In all primates, more so than in other more precocial mammalian species, the mother-infant relationship sets the stage for learning, after which other members of the social group continue the socialization process.

Before the importance of social learning was understood, primate learning tests had only been carried out in the laboratory. Although this research provided valuable insights into the mental capabilities of primates, the manner in which they learned to accomplish tasks was very different from the learning that takes place in a natural social group. One of the main centers for the study of primate learning was the Wisconson Regional Primate Laboratory, where the Wisconson General Test Apparatus (W.G.T.A.), developed by Dr. Harry Harlow, provided an efficient way of presenting discrimination problems to the animals. This invention revolutionized intelligence testing in animals, as it allowed researchers to administer a number of standardized tests in the shortest possible time. Harlow found that monkeys developed strategies allowing them to solve related problems, and that they "learn to learn" (Harlow, 1949). The W.G.T.A. subsequently became the apparatus of choice in psychology laboratories for the application of learning tests to animals and human subjects.

Learning is usually defined as a modification of behavior resulting from exposure to an outside event or from experience and practice (Drickamer and Vessey, 1982). Since it is difficult to determine if learning has taken place when there is no apparent change in behavior, in the wild one must wait for the evidence of learning to occur naturally. In the laboratory, tests can be given to indicate what has been learned.

There are three ways in which social learning in primates takes place:

1. **Observational learning** - Primates learn by watching their mothers and other group members. Young primates are ardent observers. They watch their mothers closely, gaining critical knowledge for survival: what to eat, what to fear, their social rank and appropriate social responses (see Figures 13.1 and 13.2). When infants become independent of their mothers, kin and other group members provide the role models for further learning.

2. **Social facilitation** - Research has shown that animals will increase the use of an established behavior pattern when others are performing the behavior. The observation of others actively engaged in a particular behavior appears to stimulate an individual to join in the activity (Drickamer and Vessey, 1982).

3. **Imitation** - Young primates often copy behaviors that are not normally part of their behavioral repertoire upon observing the behavior pattern in others.

Figure 13.1   An infant Japanese macaque watches his mother eat corn.

Figure 13.2   The infant has learned that it is safe to eat the corn.

Unfortunately, much of the research on primate learning is still carried out in psychology laboratories where the learning takes place in isolation, making it is difficult to say how much these studies reflect the true learning patterns in primates. In the wild, the socializing agents who help to assimilate the young into the group play an active role in the social learning of primates. Young primates who live in large multi-male groups learn easily from their peers and kin, while infants growing up in monogamous family groups acquire vital knowledge from both parents. Learning in a social context, particularly in the wild, is difficult to investigate. However, an interesting experiment in natural learning happened almost by accident (Kawai, 1965). Japanese primatologists, who were studying a provisioned troop of Japanese macaques on the Island of Koshima, noticed that an 18 month old female named Imo was washing her sweet potatoes before eating them, presumably to remove the grit. Subsequently, most of the other group members, with the exception of some adult males, began washing their potatoes. Her peers were the first to imitate Imo's behavior, followed by her mother and her other female kin. Imo also invented the use of the placer technique to remove the sand from the grain that had been distributed on the beach by the primatologists. By tossing handfuls of grain and sand into the water, she was able to scoop up the grain that had floated to the top. This new behavior pattern was taken up by group members in the same order as before, and both of the innovations became part of the troop's behavioral traditions. Although potato-washing has since been observed in other Japanese macaque groups, the placer technique for removing the sand from grain has remained unique to the Koshima troop (Kawai, 1965).

The Japanese scientists perceived a regular learning pattern in Japanese macaques and established basic rules that seemed to accompany the acquisition of new behaviors. It was clear from the Koshima experiment that animals learned easily from their peers. Although it was common knowledge that the young learned readily from their mothers, further research on other Japanese macaque groups indicated that learning new behavior patterns moves quickly from older to younger animals, and that low-ranking animals learn easily from dominant animals. The animals who approach and handle new objects in innovative ways are usually the semi-dependent young. These are the group members who accept new inventions readily and serve to perpetuate most of the useful innovations (Jolly, 1985). Although the more conservative mature males may not invent new techniques, they learned how to make use of the innovations provided by the young. Japanese primatologists regard the older monkeys as repositories of the knowledge necessary for group survival. These are the animals who can call upon their long-term memory for alternate sources of food or water in lean years (Nishida, 1987).

Nishida (1987) made a distinction between knowledge that is learned and knowledge in the form of behavior patterns that can be culturally transmitted. Knowledge that is learned, such as recognizing kin, knowing its place in the dominance rank and distinguishing group members from strangers, does not lend itself to cultural transmission because it dies with the individual.

Culturally-transmitted behaviors are taken up by group members and eventually become a group tradition that is perpetuated from generation to generation, like the potato-washing behavior in the Koshima group of Japanese macaques. These behaviors can be transferred to other groups of conspecifics by the males who emigrate from their natal troop and join another society. From further investigations of social learning, Japanese primatologists developed a "cultural transmission model" to describe the exchange of behaviors between groups of Japanese macaques. The model is based on the observation that a new behavior exhibited by an adult male quickly spreads throughout the group (Yamada, 1957). The transfer of adult males may be a way of integrating the groups behaviorally as well as genetically. When a male moves and begins to mate, his genes are introduced into the gene pool of the new group, and as he begins to interact with group members, any of his unique behavior patterns will be quickly transmitted. If culture is defined as a system of shared behaviors within a society that are passed on from generation to generation, then it could be said that there is a great deal of cultural diversity between groups of Japanese macaques, in courting behaviors, homosexual behavior, swimming, washing potatoes, manipulating stones and other behavior patterns. The use of the term "culture" by the Japanese primatologists in reference to the behaviors of their monkeys was not well received by their Western counterparts, so by way of a compromise the terms "protoculture" and "preculture" have been adopted (Imanishi and Nishimura, 1973).

164

These intergroup behavioral differences are not confined to Japanese macaques. Tomasello (1990) reported the cultural transmission of tool use and communicatory signals in chimpanzees in West and East Africa. Although there is some variation in termiting and anting techniques in the chimpanzees in three sites in Tanzania, East Africa, they all fashion sticks to probe for the insects. In contrast, West African chimpanzees crack open the termite nests and scoop up the insects with their hands. The nut-cracking techniques also differ from West to East African chimpanzees, with the former transporting large nuts to a a suitable smooth root, where they can be cracked with clubs or rocks, while the latter chimpanzees simply open the nuts by smashing them on the ground. Although the agonistic and sexual displays of chimpanzees are relatively species-specific, Goodall (1968) and Nishida (1970) reported a number of aggressive, submissive, affiliative and courtship behaviors that were unique to the chimpanzees at Gombe and Mahale Mountain respectively.

Learning and Deprivation Studies

The deprivation experiment has been the traditional method of distinguishing innate from learned behaviors, whereby an infant is raised in social isolation, creating a situation in which learning is impossible. Inherited factors are said to be the basis of any species-typical behavior, facial expression or gesture that is observed in the isolate-reared infant. Although many laboratory studies have been conducted over the years on young primates raised in varying degrees of isolation, the most widely publicized deprivation experiments were those of Dr. H. Harlow, at the University of Wisconson Primate Laboratory (Harlow and Harlow, 1962). Harlow was primarily interested in the investigation of the mother-infant affectational system which involved separating infant rhesus monkeys from their mothers' at birth and raising them with inanimate surrogate mothers. When he noticed that the motherless monkeys exhibited a number of behavioral abnormalities, he became interested in researching the effect of social isolation on primate learning. The deprivation experiments were carried out in several different social environments, ranging from total isolation to a "normal" social group. The following are examples of the types of rearing conditions that have been used in deprivation experiments (Sackett, 1974):

1. **Total Isolation** - The infants were separated from their mothers at birth and raised in a cage with solid metal walls that prevented physical, visual or auditory contact with other animals.

2. **Partial Isolation** - The infants were separated from their mothers and raised in bare mesh cages which allowed them to see, hear and smell the other monkeys but prevented physical contact with other individuals.

3. **Surrogate-reared** - The infants were separated from their mothers and raised with a variety of artificial wire-mesh surrogate mothers, some cloth-covered and some with a source of milk.

4. **Peer-only Reared** - Infants born around the same time were separated from their mothers and raised together in a wire cage.

5.   **Mother-only reared** - Infants were raised in a small cage with their mothers.

6.   **Mother and Peer-reared** - Four mother-infant pairs lived in wire-cages surrounding a central play pen that could only be accessed by the infants at specific times. This rearing environment was considered to be socially enriched since it approximated the normal rhesus social group.

The severity of the infants' behavioral abnormalities depended to some extent on the degree of isolation and the length of time spent in the isolation environment. Essentially, all of the animals exhibited abnormal behavior patterns, which Mason (1968) referred to as deprivation syndrome. The young monkeys exhibited a number of disturbance behaviors which included self-clutching, rocking and other repetitive actions. They not only had severe motivational problems and were either excessively aggressive or excessively fearful, but were totally inadequate in social situations. Although the monkeys had the urge to mate, they were unable to coordinate the motor actions necessary for copulation. Artificially-inseminated females who bore infants had no concept of maternal behavior and either abused their infants or ignored them completely. Although the infants developed some of the typical primate facial expressions, such as threats, they were unable to use them in the appropriate context (Fedigan, 1992).

Although Harlow's methods were extreme, his research did illustrate graphically the importance of social learning and the critical role of the mother-infant bond in the normal development of a young primate. It was apparent that basic behavior patterns, such as mothering and mating, were not instinctive and had to be learned in a social context. However, the research never achieved its original goal of distinguishing innate from learned behaviors. The young rhesus monkeys, who were deprived of both social and sensory stimulation and raised under stressful conditions, had such gross behavioral abnormalities that it was difficult to separate the "typical" monkey behaviors. Although this early research did provide some insights into the importance of social learning, the severity of the isolation was hardly justified (Fedigan, 1992). Jolly (1985:316) echoed the feelings of many when she wrote: "And once the first dozen rhesus had been reared in isolation and the first surprise was over, did the later results justify the misery knowingly inflicted?" Less severe separation studies have been carried out since the 1970s, where the monkeys are raised in stable social groups and the degree of isolation is controlled. These studies have demonstrated a number of interesting aspects of the mother-infant relationship and the effects of short-term separations on the infants. For example, it has been shown that the effects of maternal separation are less disturbing for the infant if it is removed from the mother, rather than taking the mother away from the infant. The results of of these studies have been found to be applicable to human mothers and infants, in cases where a short-term separation is necessary (Hinde and Spencer-Booth, 1971).

Subsequent research on the mother-infant relationship in primate species other than rhesus macaques has indicated that the reaction of the infants to separation from their mothers is variable, depending on the social patterns of the group to which they belong. Kaufman and Rosenbloom's (1969) investigation of species differences in allomothering, discussed in Chapter 11, clearly showed that infants were unaffected by maternal separation in species where the mothers were permissive with their infants.

Summary

The scientific community has generally accepted the fact that behavior is shaped by a combination of innate characteristics and learned experiences from the social and physical environment. The genotype sets the biological limits for a particular behavior to occur, however, as Jolly stated, "the relative importance of learning and instinct depends on the range of environmental conditions sufficient and necessary for a behavioral trait to appear" (Jolly, 1985:190). Primates are social animals and all of their learning takes place in a social context. Although laboratory studies have provided insights into primate learning abilities, true learning patterns must be investigated in a natural setting. Young primates learn by observing their mothers, kin and peers, by social facilitation and by imitating the behaviors that they have so carefully watched. Research has shown that new behavior patterns can be "culturally" transmitted throughout a primate group and eventually become a tradition in that society. In some cases, particularly in species where the males transfer between groups, unique behavior patterns can be transmitted to a new group of conspecifics and become part of their social tradition.

Chapter 14

Communication

Recent advances in technology, along with a steadily-increasing body of data, have led us to re-evaluate many of our basic assumptions about the nature of primate communication. It was once assumed that nonhuman primates used a limited repertoire of signals. Now it is clear that primate communication systems are extremely complex and are integrated into every facet of their daily lives (Zeller, 1987). Primatologists are now able to recognize and analyze a great many signals with the help of sophisticated equipment such as camcorders, tape recorders, gas chronometers and computers, allowing more accurate predictions about the information enclosed therein.

The study of animal communication systems is important because these external signals are the manifestation of internal processes and provide us with some insight into matters that are important in their daily lives. The immediate function of communication is to disseminate information related to the daily needs of the individual and the social group: the recognition of kin, the position in the dominance rank, the location of food, the location of conspecifics, and the presence of predators. Ultimately, communication systems function to maximize reproductive success and to promote survival in the social and physical environment.

The Act of Communication

The four distinct features that can be recognized in an act of communication are the **signal,** the **motivation,** the **meaning** and the **function** (Smith, 1977).

1.   The signal is the observable action (a shriek or a threat face) that can be recorded in some way, depending on the mode of communication. The signal may be simple, composed of one behavioral unit, or a complex communication made up of a number of elements which can be described as one signal or broken down into the smaller units, depending on the focus of the observer. A **display** is a ritualized, stereotyped communicative act which is much more elaborate than a mere signal since it often incorporates many gestures and signals. The spectacular sight of male chimpanzees with hair bristling, charging around the forest throwing foliage in all directions, or the silverback gorilla standing erect, beating his chest, is indicative of the highly-charged emotional situations that evoke primate displays.

169

2. The motivation behind the communication can often be established by observing the actions that accompany the signal. A growl given in conjunction with a bite or a cuff clearly indicates an aggressive motive, while a growl accompanied by a play face suggests a totally different motivation. Inferences about the motivation behind a signal must be based on the relationship between the signal and the context in which it is given. If the signal occurs in the same form in similar contexts, then the assumption about motivation is likely to be reliable. Since it is often difficult to avoid **anthropomorphism** and ascribe signals to human feelings such as "hate" or "love," the observer must take care to record motivations in objective terms such as aggression and affiliation.

3. The meaning of the signal can be determined by the reaction of the receiver. The context of the communicative act is particularly important in understanding the meaning of the signal because individuals may react very differently, depending on the situation. An animal who challenges an aggressor in the company of kin may avoid such a confrontation when alone or with unrelated animals.

4. The function of an act of communication is the adaptive significance or the ultimate effect of the signal.

The high-pitched scream of a vervet on the African savanna that causes the other group members to look up and run to the base of a tree is a good example of a communicative act encompassing all of these four features. The signal is the scream, which indicates that the motivation is fear when given in the context of a crowned eagle soaring overhead. The meaning of the scream, judging by the reaction of the group members, is that an aerial predator is nearby and the function of the signal is to protect the group members and kin from predators (Seyfarth et al., 1980).

Modes of Communication

Although most primate signals are multi-modal, primate communication systems are usually described in four separate modes: **olfactory, tactile, visual** and **vocal.**

Olfactory Communication

Olfactory communication is carried out by chemicals given off by the sender. These chemicals are transmitted through the nose to chemoreceptors in the brain of the receiver, where the messages are interpreted. The volatile chemical communicators, which are called pheremones, are produced in specialized glands found in many primate species, and transmitted alone or mixed with urine or feces. Since an observer is unable to experience the chemical message, olfactory communication is extremely difficult to study and to understand.

Olfactory communication is of primary importance to a number of solitary nocturnal prosimian species because the messages persist over time and can be received long after they're given. The moist prosimian nose, which is connected to the upper lip, is well-adapted to picking up olfactory signals, but results in a relatively expressionless face. Fortunately, the inability to communicate by means of facial expresssions is unimportant in a nocturnal creature. Although olfactory communication possibly evolved in the ancestral nocturnal primates, it is used extensively by diurnal group living species, such as ring-tailed lemurs, callitrichids and some cebids, indicating its wide versatility in the transmission of information (Zeller, 1987).

There are two types of chemical communicators that vary in their mode of action. **Releasor pheremones** elicit an immediate effect, while **primer pheremones** produce a delayed reaction and alter the state of the receiver over time. It is the releasor pheremones, produced in the wrist glands of male ring-tailed lemurs that stimulate the "stink fights" observed in competitions for estrous females. Primer pheremones are assumed to be responsible for physiological phenomena, such as the suppression of reproduction in young marmosets and the effect of estrous females on the production of testosterone in males.

Scent marking, which is common to many mammalian species, is carried out in a number of ways and serves a variety of purposes in nonhuman primate communication (see Figure 14.1). The scent can be dispersed into the air by the movements of the animal, or it can be deposited on the substrate in a precise, stereotyped manner at a particular time and place. Squirrel monkeys drip urine on their hands and feet and leave messages in their sticky footprints on the branches. In contrast, the urine rub of the cebus monkeys, is used to mark their bodies rather than the **substrate**. Prosimian species not only use scent marks to define their territories, warning off other groups of conspecifics, but also

Figure 14.1   Female brown lemur scent marking
(Courtesy Ruben Kaufman).

to attract members of their own group by advertising their dominance status, their sexual receptivity, or just their general wherabouts. Some species use olfactory signals in aggressive encounters over competition for dominance or mates. A prime example of this is the tail-waving display, characterizing the "stink fights" of the male ring-tailed lemurs. There is evidence that brown lemurs and tamarins are not only able to recognize age and sex by scent marks, but can identify individuals by their smell (Zeller, 1987). Humans strive to remove their body odors by daily bathing and the use of deodorants, and do not generally appreciate the effect of scent on their behavior. However, extensive research on human olfaction is carried out by perfume companies who recognize that, even in human primates, olfactory communication is important and can be very lucrative.

Tactile Communication

Although the importance of tactile communication in primates is recognized, it has received the least attention in the literature because observers are unable to perceive the signal in the same mode as the recipient. Since researchers cannot experience the contact, they have to rely on the reaction elicited to interpret the signal. Tactile communication occurs in intense, intimate social interactions, such as those between a mother and her infant or between a female and her male consort. Harlow and associates recognized the importance of contact comfort in new-born primates when they found that infant rhesus monkeys preferred to spend most of their time clinging to the warm cloth-covered surrogate mother than on the wire-mesh apparatus that supplied nourishment (Harlow et al., 1963).

Grooming is the form of tactile communication that has been studied the most intensively because it appears to play such a prominent role in the daily life of many primate species. Although one of the important functions of grooming is to rid the body of dirt, flakes of dry skin and external parasites, the time devoted to it would suggest that the benefits go beyond personal hygiene. Grooming must be important in maintaining the even tenor of group life because it has been observed in so many different social contexts: mothers pacify their infants by grooming them, consort pairs groom as a prelude to mating, kinship bonds are reinforced and agonistic encounters are often followed by bouts of grooming to mend ruptured bonds (see Figure 14.2). Grooming can be used as a diversionary tactic by mothers who are attempting to wean their young, and by the offspring who are intent on suckling. It seems to lull the recipient into a state of drowsy

Figure 14.2 Grooming bout in a kin group of Japanese macaques. (Courtesy Karen Dickey.)

Figure 14.3　Being groomed appears to be a very soothing experience. (Courtesy Linda Fedigan.)

euphoria, where other matters, such as weaning or suckling are temporarily forgotten (see Figure 14.3). The opportunistic grooming of high-ranking animals is a strategy often used by subordinate individuals as a means of sharing some of the benefits of dominance. There are even instances where individuals have moved up the social ladder in a primate society by the assiduous grooming of the alpha male or female.

As with other types of communication, there is a wide interspecific variability in the style and the amount of grooming observed. Despite the obvious importance of grooming in many nonhuman primate societies, some species, such as squirrel monkeys, rarely groom each other. Monkeys and apes groom by parting the fur with their fingers and picking out the foreign particles with their hands or teeth, while prosimians use their tongues and tooth combs to groom other group members. Animals who are related or who know each other will groom each other daily, while individuals who are unfamiliar with each other must go through elaborate initiation gestures before invading the personal space of a prospective grooming partner (Jolly, 1985).

Self-grooming, like preening in birds, serves an important function in the personal hygiene of primates, but it also seems to have a calming influence on animals under tense or stressful conditions (see Figure 14.4). In captivity, primates who are isolated from grooming partners sometimes groom themselves pathologically to the point of self injury.

Primates are particularly tactile animals and have an impressive repertoire of tactile communications, aside from grooming. Although most species huddle together at night and show affiliative muzzles and pats, as well as agonistic cuffs and bites, chimpanzee greetings, with their effusive hugs and kisses, are the most varied and human-like. (Jolly, 1985).

Figure 14.4   A vervet monkey self-grooming.
(Courtesy Karen Dickey.)

Visual Communication

Visual communication systems have been more widely studied than the other modes, because the signals can be more readily observed and understood. Although visual messages can be transmitted by different physical signals, such as piloerection (hair standing on end) and the tilt of the ears, posture and facial gestures play the most important roles in visual communication in primates. The expressive faces of most anthropoidal primate species can convey messages at close-range, while the postural communication is obvious from a distance. High-ranking male macaques and baboons can be easily distinguished from the others by their confident stride and upright tails, while subordinate males assume a more submissive posture with knees bent and lowered tails (Jolly, 1985).

Visual communication serves a number of functions in nonhuman primate society, in both competitive and cooperative situations. Threats can be used to warn off an aggressor, or as in hamadryas baboon society, can be used by the male to coerce a female. Not only does the primatologist studying visual communication have to be aware of the relationship between the communicants, but must take careful note of the situation in which the message occurs (Zeller, 1987). There are numerous examples that illustrate the importance of the context of the communication in understanding a visual signal. The yawn of an adult male baboon may be interpreted as a relaxed expression during a resting period, or a tense, threatening signal in an agonistic encounter with another male. Play faces accompanying a bout of wrestling between two juveniles indicate that the potentially agonistic activity is all in fun (Jolly, 1985).

Harlow's research clearly showed that, although many gestures and facial signals appear spontaneously in young primates, the appropriate use of the signals must be learned in a social context (Harlow and Harlow, 1962). This is especially true for the ability to manipulate social relations by the judicious use of threats, displays and solicitations for help (see Figures 14.5 and 14.6). The degree of sophistication in primate facial and postural communication used in a natural setting is impressive. Research in the laboratory and in the field has shown that nonhuman primates are not only able to transmit information in the visual mode, but can suppress signals to withhold information, and are even capable of intentionally giving false messages (Menzel, 1974; Premack and Woodruff, 1978).

Color vision is important in diurnal anthropoids, particularly arboreal species whose ability to distinguish different colors of green facilitates movement through the trees. Some species of monkeys, such as the forest-dwelling guenons, use their spectacularly colored faces, coats and hind quarters to transmit signals to other group members and to accentuate differences between closely-related species. Transient non-volitional signals also serve to communicate important information. For example, the perineal swellings and reddened facial and sexual skin of estrous female baboons communicate a willingness to mate, and the temporary neo-natal coat of infant primates allows them a greater latitude in their interactions with adults.

Figure 14.5   Threat face. (Courtesy Karen Dickey.)

Figure 14.6 Fear grimace. (Courtesy Karen Dickey.)

Vocal Communication

Early primatologists, who relied on the human ear to interpret and classify the vocalizations of nonhuman primates, assumed that their repertoire of signals was extremely limited. Modern technology has revolutionized the study of animal vocal communication systems, and we now have the tape recorder to capture the calls and the **sonogram** to provide a spectrographic picture of the sound waves. Clearly, primates communicate by means of a wide range of calls, using variations in pitch and intensity to alter the message.

Although the messages are transient, the advantage of vocal communication over the other modes is its ability to catch the attention of the receiver. The long distance calls used by arboreal species are loud and low-pitched to carry through the cover of trees. The fluctuations in the temperature and air currents is a factor in the timing of these calls, and as a result, the forests ring with loud vocalizations at dawn when the sounds carry well. Sharp, short barks used as spacing mechanisms are easy to locate, while whistles or shrieks, with undefined starting and finishing points, are difficult to place and make effective warning calls. The vocal repertoire of a primate species is often closely related to their form of social organization. Although primates living in monogamous family groups tend to have a more limited vocabulary than species living in large complex societies, the mated male and female often sing complex duets. Chimpanzees, with their pant-hoots, screams and grunts, have one of the most varied vocal communicative systems of any primate species. It serves them well, as their fluid social organization involves the need to advertise fruiting trees and to enrich their frequent reunions with group members with noisy greetings. In contrast, gorillas, who rarely stray far from their social group, make do with a few soft belches (Jolly, 1985).

The study of the kinds of information transmitted vocally has proven to be an important area of research because it has provided valuable insights into the depth of social knowledge exhibited by primate species. Cheney and Seyfarth's innovative experiments on free-ranging vervets have advanced our understanding of the sophistication of primate vocal communication. They have demonstrated that wild vervets are able to identify individuals by their vocalizations, and can recognize relationships between different members of the social group. Using a hidden tape recorder to play back the screams of a two-year-old juvenile, they were able to determine by the mother's response that she recognized the screams of her offspring. The reactions of females near the mother clearly indicated that they knew the call belonged to her offspring and not their own (Cheney and Seyfarth, 1980). Cheney and Seyfarth also established the fact that vervets have at least three separate and distinct alarm calls that elicit different reactions from group members. The vervets look up and run for the cover of the trees when warned about a raptor, while a leopard alarm call sends the monkeys up into the trees. Upon hearing a snake alarm, the vervets stand up, peer at the ground and often react by mobbing the snake (Seyfarth et al., 1980). Their extensive research on vervet vocal communication has prompted this observation from Chebey and Seyfarth (1990:138): "Because nonhuman primates use vocalizations to signal about *things* (their emphasis), research on communication offers a glimpse of how they see the world."

Ape Language Experiments

The dream of communicating with animals has captured the imagination of scientists and non-academics alike. There have been a variety of reasons for this fascination, some more fanciful than others, ranging from the notion that if we could converse with animals we could find all sorts of interesting things about their philosophy of life, to scientific lines of inquiry such as: the determination of the extent to which apes acquire language, a comparison of the development of linguistic ability with that of children, the issue of the continuity of cognitive and linguistic abilities from apes to humans, and the possible application of similar techniques for teaching aphasic children to use language. Unfortunately, the results of several ape language studies have been subjected to controversy and criticism, which has tended to divert attention "from the important aspects of an ape's ability to communicate symbolically, however primitive that ability may be." (Terrace, 1985:1011).

In the early ape language studies, the emphasis was on teaching apes to use human verbal language based on the assumption that apes are unable to speak because they have not been exposed to human speech. The first well-documented study was conducted by the Kellogs (1933), a husband and wife psychologist team who raised Gua, a female chimpanzee, with their son Donald, in order to compare the motor and mental development of the ape and the child. Since young apes mature more rapidly than children, Gua outstripped Donald in locomotor and specially-adapted intelligence tests until they were over a year old. The experiment was terminated at 18 months when it was obvious that the boy was acquiring too many ape-like traits, and at the end of the study, Donald could speak, but Gua had not uttered a word. The experiments of Keith and Kathy Hayes (1951) with Vicki, another female chimpanzee, were slightly more fruitful. After six years of intensive training, her verbal repertoire consisted of four barely-understandable words. Although the primary objective of these experiments was never achieved, it was obvious that the apes were very adept at non-verbal tests. Vicki could understand spoken words and could sort a number of objects by size, shape and color (Hayes, 1951), while Gua had no difficulty mastering intelligence tests adapted for apes and young children (Kellog, 1933). The inability of the ape subjects to speak was a great disappointment to their mentors, and it was suggested that apes either do not have adequate cerebral structures or the appropriate vocal apparatus for speech (Ristau and Robbins, 1982).

Experiments Using American Sign Language

Beatrice and Allen Gardner (1969), psychologists at the University of Nevada, studied the films of Gua, Vicki and wild chimpanzees, and were convinced that the apes could master a gestural form of communication. In 1962, they developed a simplified version of American Sign Language (ASL) and a technique suitable for teaching it to Washoe, a young female chimpanzee. "Experiment Washoe" was designed to test the ability of a chimpanzee to use a sign language in two-way communication with humans, and as such, it was an unqualified success. She learned signs quickly and was soon able to string them together in two-word phrases. Washoe was housed and trained in a trailer, complete with a number of stimulating objects, and was tutored by the Gardners and others who had recently learned ASL. Her progress was documented by means of written reports, tape recordings and films, and by the time she was four years old, Washoe was able to use 132 signs reliably. It was claimed that she had mastered the appropriate use of words and word combinations, and that she showed spontaneity and creativity in her word production (Gardner and Gardner, 1969). This experiment was important, not only because it paved the way for all similar studies which used the same or slightly modified techniques, but because it emphasized the need for studies investigating the acquisition of language in children (Ristau and Robbins, 1982).

Surprisingly, their fellow psychologists remained skeptical about the success of the Gardners' research on the grounds that it lacked proper controls and that the reports of the trainers were biased. Although it was evident that Washoe had learned a number of ASL signs, there was a major effort to discredit ASL as a true language. This situation has since been corrected, and now ASL is considered to have all of the attributes of language.

Project Washoe was terminated after six years and the Gardners embarked on a second project with the objective of investigating the development of ASL in infant chimpanzees and comparing it with that of children. In this study, they used trainers who were fluent in ASL in an effort to improve the accuracy of the signing, and tried to restrict the use of verbal language with their subjects. They acquired four infant chimpanzees and began working with them shortly after birth, observing the development of signing ability. The first sign appeared at three months, compared to four months in human infants, and the results suggested to the Gardners that the acquisition of sign language follows a similar pattern to that observed in human infants (Gardner and Gardner, 1979).

Similar studies, which were initiated in the 1970s are still continuing, indicating the long-term commitment that has characterized these studies. Roger Fouts, a former student of the Gardner's. initially worked with six young chimpanzees to investigate the possibility of communication between ASL-using apes. He reported that they had learned from 40 to 70 signs after two to four years of training and showed spontaneous use of signs and novel word combinations in their communications between each other (Fouts, 1977). Fout's most recent project is concerned with the ability of a young chimpanzee to learn signing from its mother, using Washoe and her adopted son, Loulis, as the subjects. Loulis proved to be an apt pupil and has learned to sign from Washoe and the other chimpanzees without any help from the trainers, suggesting that human intervention may not be necessary for the acquisition of signs (Fouts, 1983).

Francine Patterson's study is the only one to date that has used gorillas as the subjects of an ape language experiment. There had been a bias against using gorillas in any type of psychological testing because chimpanzees were thought to much more gregarious and intelligent. The success of Patterson's experiment has dispelled this notion and has shown that, with the proper level of stimulation and a stress-free environment, gorillas are able to learn equally as well as chimpanzees. She began teaching ASL to Koko, a female lowland gorilla, in 1972, and has since added Michael, a young male, to her study. The gorillas are housed and trained in a wooded area of Woodside, California, allowing them limited access to the outdoors. The trailer provides a relaxed atmosphere for the gorilla's training program, which includes reviews, exercise and play (Patterson, 1981). According to Patterson, Koko has mastered well over 500 signs, which she uses in many creative ways, showing the ability to deceive, joke, insult, use novel word combinations and recognize herself in the mirror. Patterson is continuing to work with Michael and Koko, and her future plans include mating them and investigating the transmission of language from parents to offspring.

Lyn Miles used a somewhat different perspective in her study of sign language acquisition in Chantek, a male orangutan. Her objective was to investigate the development of the cognitive processes that are basic to the use of symbolic behavior in communication, rather than language acquisition per se (Miles, 1990). Chantek's living and training environment was a trailer equipped with a variety of toys and playthings in the University of Tennessee campus. Since an arboreal environment is important for an orangutan, the trailer was situated in a treed courtyard fitted with ropes and a jungle gym. Miles, like Patterson, tried to keep the learning sessions as relaxed as possible, on the premise that the lack of pressure would enhance the motivation to learn. Chantek proved to be a very apt pupil and was able to use 140 signs effectively by 1986, when the study was discontinued. Miles argued that, although the signing of the orangutan was slower and more deliberate, he was able to initiate conversations, sign spontaneously and was generally more articulate than the signing chimpanzees (Miles, 1983).

Artificial Language Experiments

The experiments designed to study linguistic capabilities of apes have not been confined to the use of gestural systems as a means of communication. Duane and Sue-Savage Rumbaugh initiated project Lana (Language Analog Project), in which they trained Lana, a female chimpanzee, to communicate by means of geometrical lexigrams on a computer keyboard. They named the artificial computer language "Yerkish" in honor of Robert Yerkes. Their primary objectives were to investigate the communicative ability of apes and to test this approach as a possible method of teaching **aphasic** and retarded children to use language. They also felt that the use of symbols on a computer keyboard would preclude the over-interpretation of ambiguous ASL signs and eliminate the possibility of cueing. Lana was trained to form associations between the lexigrams and the objects they represented and then to use phrases to convey message to the trainers. She learned the symbols for 75 items and was able to communicate requests, such as "Please machine give M and M," to the trainer (Rumbaugh and Gill, 1977).

The Rumbaugh's second experiment involved the investigation of communication and social cooperation between chimpanzees. The subjects were Sherman and Austin, who were trained to use the lexigrams on the computer to ask each other for the appropriate tools to retrieve food from baited boxes, and then to share the food. Although they learned the symbols for over 50 items and were able to use the lexigrams effectively to communicate their requests, it was difficult to train them to share the food (Savage-Rumbaugh et al., 1978).

Sue Savage-Rumbaugh is continuing her research at the Yerkes Language Research Center at Georgia State University in Atlanta with common and pygmy chimpanzees. Her brightest pupil is Kanzi, a male pygmy chimpanzee, who surprised the researchers by learning lexigrams by observation, without the tedious step by step training used for the common chimpanzees. The gestures that he has incorporated into the lexicon system allow him a much greater flexibility in his communications, to the point where the lexicon board is portable and no longer connected to the computer. The action-based rules used by Kanzi to order the elements in his communications, could be construed as grammar or at least a proto-grammar. Since early humans had to begin by inventing their own grammar, Greenfield and Savage-Rumbaugh (1990) feel that Kanzi's invention of grammatical rules is evidence of a continuity in language abilities from apes to humans.

Another artificial language project was initiated by Anne and David Premack in 1964, at about the same time that the Gardners began working with Washoe. They were interested in breaking language down into its basic components and testing the ability of apes to master the use of quantifiers, negatives and interogatives. Sarah, perhaps the most test-wise chimpanzee to date, was one of their four chimpanzee subjects. The Premacks taught the apes to communicate by using magnetized plastic chips as symbols of words that could be arranged into sentences. They were also interested in testing the ability of their subjects to master concepts such as same/different and color/shape. When the study was terminated, Sarah had learned 130 symbols and could create hierarchically organized sentences containing all of the linguistic components (Premack and Premack, 1981). Premack's subsequent research, which has been directed towards the investigation of mental states such as intentions, beliefs an deception in apes, will be discussed in Chapter 15.

The Controversy

Clearly, the emphasis in ape language studies has changed over time, from teaching apes to speak, to communicating with them using gestural and artificial systems, and more recently to investigating mental constructs used in social interactions with their own species. The change in focus relates in part to problems encountered in adequately defining language, which has presented difficulties in assessing the linguistic skill demonstrated by the apes (Ristau and Robbins, 1983).

Over the years, there has been a lively and often bitter controversy over the claims made concerning the apes' mastery of language. One of the most vocal critics of the ape language studies has been Herbert Terrace, a psychologist at Columbia University, who conducted his own ape language project in the early 1970s with Nim Chimpsky, a male chimpanzee. Terrace was particularly interested in the ability of an ape to use grammatical rules in communication. Nim was raised in a home-like atmosphere, but was trained in a bare, featureless classroom, designed to provide few distractions and to facilitate the observation and the documentation of Nim's progress. Over the 44 month study, Terrace used 60 non-fluent ASL trainers, many of whom had little contact with Nim outside of the formal training sessions. After the completion of his study, although Nim had mastered 125 signs, Terrace was skeptical of the ape's linguistic ability. He carried out an exhaustive analysis of the data, looking at the structure and the mean length of the utterances, spontaneous signing and the novel use of signs. His findings formed the basis of his critical attitude towards the claims of the other researchers. Terrace was concerned about the limits of the apes' linguistic abilities, specifically the capacity to use grammar and understand the meaning of their utterances. His detailed analysis of the data from the Nim Project and the films of Washoe and Koko convinced him that the apes were using strings of words with no true grammatical structure or conversational rules. They were merely memorizing the ASL signs to get rewards from the trainers. Terrace contended that all ape language studies showed similar tendencies, and that the impressive claims of the researchers were the result of loose definitions of spontaneity and novel word combinations, overinterpretation by over-eager trainers and cueing (Terrace, 1983).

These criticisms were noted by funding agencies and money was no longer available for future ape-language research. Not surprisingly, this provoked a strong reaction from the other researchers, who blamed Nim's deficits on Terrace's training regime. They argued that the sterile classroom atmosphere, the drill-like training sessions and the use of 60 trainers would suppress the motivation for creativity and spontaneity in any individual. Most of the other studies had reported that the spontaneous and innovative utterances had been observed in casual, relaxed situations, and that the signing tended to become perfunctionary with constant repetition. Given that chimpanzees are gregarious social animals who become attached to their trainers, it was argued that Terrace had ignored the social aspects of language (Ristau and Robbins, 1982). Although Terrace agreed that some of his training practices were less than perfect, he continued to argue that he had seen no evidence that apes can master the conversational, semantic and syntactic aspects of language. Perhaps the evidence from the continuing research with Kanzi and other pygmy chimpanzees, will convince sceptics about the abilities of apes to communicate symbolically using the systems that we have invented.

Another criticism that has been levelled at these experiments has centered on the possibility that apes are reacting to subtle, inadvertent cues of their trainers. The ASL projects are particularly susceptible to cueing because they take place in a relaxed atmosphere in close contact with their trainers, who are anxious for their subjects to perform well (Seboek and Umiker-Seboek, 1983). Fouts (1983) and Miles (1983) agree that cueing is difficult to avoid, but suggest that all conversations contain non-verbal cues that provide important information in communication. Miles also argued that the construction of an ASL sign is too complex to be cued by a simple "on-off" signal, and that the apes' ability to pick up subtle cues indicates that they are exceptionally perceptive.

Eugene Linden, a journalist who was closely connected with ape language studies for many years, wrote that the jury is still out as to whether the ASL signing apes demonstrate a capacity for language, and that "The question is insurmountable until science decided what language is," (Linden, 1986:24). It is obvious that, while apes can communicate with humans using systems that we have taught them, they are not able to achieve the linguistic capability of human adults or children. However, one could ask if it really matters if apes are able to use syntax or grammatical rules, or if they understand the meanings of symbols in the same way we do. It is perhaps more important to understand the lesson that these interesting and valuable experiments have taught us: Apes have prodigious cognitive capacities, allowing them to cope successfully with the uncertainties of life in their physical and social environments. They have demonstrated excellent memories, allowing them to learn gestures and artificial symbols readily and are adept at using them to communicate their needs. They have learned to communicate with each other using our systems and have shown that the young can learn our signals without human intervention. Having read Jane Goodall's accounts of the sophistication of chimpanzee communication in the wild, surely the results of these studies have only confirmed what we have suspected about the cognitive ability of apes.

Summary

The study of primate communication systems has allowed us a glimpse into their most intimate moments, as well as helping us to identify many of their everyday concerns connected with group living. In order to better understand the messages being sent, the communicative act is often viewed as being made up of the following four components: the signal, the motivation of the sender, the meaning to the recipient as judged by the reaction elicited, and the ultimate or evolutionary function. The four modes of comunication used by nonhuman primates provide different advantages. Olfactory signals, in the form of the scent marks used by prosimians and some New World monkeys, are long-lasting, while the transient vocal signals function to alert group members. Visual messages, are usually directed to a specific individual or group, and since they are often obvious, they are more easily detected by the researcher. Tactile messages, which are the most difficult to understand because they cannot be experienced by the observer, often take place in intense, intimate interactions. Many primate communications are multi-modal, using one or more of the channels available to them.

Traditionally, scientists have been interested in communicating with animals. Laboratory experiments have proven that the great apes are able to master gestural and artificial forms of language, and use them to communicate with their trainers, and even with each other. Not only have these studies illustrated the impressive cognitive capacities of the apes, but have been instrumental in stimulating research into the way in which children acquire language.

Chapter 15

Primate Cognition

An Historical View of Animal Cognition Studies

The subject of animal cognition or animal intelligence has fascinated people for generations. Some, like Aristotle have totally discarded the idea of animals, believing that verbal language is a prerequisite for thought (Griffin, 1978). Others, such as the Roman naturalist Pliny, have argued strongly for intellectual capabilities in creatures other than man. The first serious work attempting to analyze mental states in animals was <u>Animal Intelligence</u>, published by George Romanes in 1882. He was greatly influenced by the writings of Darwin and proposed an evolutionary theory suggesting a continuity of mental processes from lower to higher forms of life. His aim was to investigate the extent of animal cognitive processes and compare them with those of humans. He proposed that studies in a new field of science, which he called comparative psychology, could discover evolutionary trends in mental abilities. Since there is no way of proving the existence of mind in others, Romanes suggested that the only way to study mental phenomena was to observe behavior. He reasoned that similarities between animal and human behavior indicated similar thought processes. Unfortunately, the only available data to support his theory was anecdotal material compiled from observations of the behavior of a number of animals, including monkeys (Romanes, 1882).

C. Lloyd Morgan countered Romanes' inferential method of analysis with <u>An Introduction to Comparative Psychology</u> (1896), in which he tried to raise the new discipline into a more objective realm. His mechanistic approach to comparative psychology was advanced by influential psychologists like Thorndike, Watson and Skinner, and the rigorous school of behaviorism became firmly established in United States (Wasserman, 1984). The basic premise of behaviorism was that the stimulus event, the response and the change in the relationship between them can be observed. Thus, behavior could be manipulated, controlled and predicted, since it was said to be the product of learned responses to simple and complex stimuli. Morgan's canon of parsimony, which dictated that the simplest explanations for observed behaviors are preferable to those that use complex interpretations, became an important tenet of the school of behaviorism. This effectively ruled out the concept of consciousness in animal and human behavioral studies, since it would be not be parsimonious to credit a behavior to a complex mental process when it could be explained by a stimulus-response mechanism (Griffin, 1978). Griffin, an ardent advocate of animal consciousness, suggested that many animal behaviorists have been concerned with observable behavior and have been inclined to ignore mental processes because they cannot be directly observed. He proposed that a new discipline, **cognitive ethology,** be designed to deal with the analysis of animal behavior within a cognitive paradigm (Griffin, 1978).

In psychology, the stimulus-response paradigm has been severely challenged by its inability to explain behavioral phenomena, particularly those dealing with memory in humans and animals. On the other hand, there is ample evidence for the impressive mental capacities of nonhuman primates, and the study of animal cognition has achieved respectability. Comparative psychologists have shown that primates are generally adept at problem-solving and mastering a whole battery of discrimination tests in the laboratory, while primatologists working the field report the use of sophisticated social skills in the daily lives of primates (Kummer, 1982; Goodall, 1985; Strum, 1985).

Laboratory Intelligence Tests

Kluver (1937) was one of the earliest scientists to test the learning abilities of monkeys and apes in a variety of different tasks, designed with the specific purpose of determining the mechanisms by which the tasks were accomplished. Harlow's invention of the Wisconsin General Test Apparatus streamlined the application of learning tests by providing a standardized, efficient way of determining the ability to discriminate a variety of objects. As his rhesus monkey subjects mastered the discrimination test for one set of objects, different sets were used until several hundred tests had been administered. The rate of learning increased dramatically as the animals became more experienced, which prompted Harlow (1949) to suggest that the monkeys formed learning sets and were, in effect, learning to learn. At first, it was thought that that the ability to learn from experience increased with the evolutionary complexity of the organism and that monkeys would outscore rats, and apes would outscore monkeys. However, when further testing showed that rats, cats and birds could become as adept at learning sets as monkeys, it became clear that species differences in learning depended on their adaptations to the environment and that one species was not necessarily "more intelligent" than the other. Although monkeys are better than rats at visual discrimination tests, rats outperform monkeys on the discrimination of odors (Tarpy, 1982).

More sophisticated tests have indicated that nonhuman primates are able to master tasks that require the ability to use abstract concepts, such as **analogical reasoning** and **transitivity.** Sarah, Premack's test-wise chimpanzee (see Chapter 14), has demonstrated the ability reason analogically in two different tests:

1. She was able to understand the functional relationship between sets of objects, such as equating a key/lock combination with can-opener/can on the premise that one is used to open the other, or equating scissors and paper with a knife and an apple.

2. She could judge the equivalence of relationships by completing the analogy A is to A' as B is to B', using geometric symbols differing in color, shape and size.

In a transitive series of 3 objects, if A is bigger than B and B is bigger than C, then it follows that A is bigger than C. Gillan (1981) used a complex task to test the ability of chimpanzees to understand the concept of transitivity by using the relationship of "more food than" between three adjacent pairs of colored cups. one containing food and one empty. Cup B contained food when paired with A, but was empty when paired with C, C was full when paired with B but empty with D and so on. One subject in particular could readily predict the location of the food, indicating that she could infer that the relationship of "more food than" held for non-adjacent pairs. McGonicle and Chalmers (1977) have even been able to show that squirrel monkeys can master transitivity tasks.

Woodruff and Premack (1978) tested mental states such as intentions, beliefs and deception in Sarah and other chimpanzees in the following ways:

1. Sarah's comprehension of the intentions in others was tested by showing her video tapes of actors confronted with problem situations, such as reaching for an object that was too far away. When given a forced choice between two photographs, only one of which depicted the correct solution to the problem (a rake to move the object closer), her perfect performance indicated that Sarah understood the intention of the actor. They chose a similar test to demonstrate that Sarah's preference for certain individuals would color her decision as to the appropriate solution for problems. Sarah was shown a favorite trainer and one whom she disliked in problem situations, and was asked to choose from three photographs, one depicting the correct solution, one depicting a catastrophic solution and one depicting an irrelevant solution. In most cases, her choices reflected her feelings towards the actor, since she picked the correct solution for her friend and the catastrophe for her enemy. It is interesting to note that she never picked the irrelevant solution.

2. The ability to direct and understand misinformation was tested by using a cooperative and an uncooperative trainer in situations involving the acquisition of food. The "mean" trainer would always attempt to direct Sarah to the empty bin. If he was directed by Sarah to the bin containing food, he would greedily eat it all. The "nice" trainer directed Sarah to the correct bin and shared the food with her when she showed him where it was hidden. Sarah quickly began to act differently towards the trainers, and when confronted by the "bad" actor disregarded his attempts to direct her to the wrong bin, and deliberately withheld signals directing him to the correct bin. Her performance demonstrated that she was able to understand deception and deliberately withhold or give misinformation.

Problem-solving tasks are more open-ended than discrimination or concept learning tests because the subject has to figure out a way to complete the task rather than make a correct choice of two or more alternatives (Fragaszy, 1985). Although chimpanzees and orangutans have proven to be particularly proficient in terms of their ability to solve problems, cebus monkeys are exceptionally adept at mastering puzzles that require manipulative ability. It should not be surprising to note that the abilities demonstrated by monkeys and apes in laboratory tests have adaptive significance in their natural environments. The excellent visual acuity and good memories that help them to master discrimination tests are useful in finding and remembering sources of food and water. The ability to categorize items and understand concepts such as transitivity are useful in a physical and a social sense, enabling them not only to distinguish between different types of food items and different groups of conspecifics, but also to readily understand their place in the dominance hierarchy. The ability to reason analogically is useful in generalizing the results of past experiences to novel but similar events. It has been suggested that differences in the capacities across species can be understood by viewing them in terms of their usefulness in coping successfully in a physical and social sense within a given environment.

Tool Use and Intelligence

Although we all understand what tool use implies, there is no universally accepted definition of the term. Some make no distinction between simple object manipulation and intelligent tool use, while others, such as Beck (1980), who reviewed most known cases of tool use in wild animals, arrived at at a very specific, if lengthy, description of what tool use entails. He suggested that tool user had to employ an unattached object to alter another object, organism or the user itself in some way by the proper direction of the tool. The tool use that has been reported in many species of monkeys and apes usually occurs in three main contexts: as weapons, in the acquisition of food, or for bodily care (Jolly, 1985; Essock-Vitale and Seyfarth, 1987).

Weapons

Many arboreal species merely drop branches and foliage on humans and other terrestrial intruders, while cebus monkeys and chimpanzees have been observed aiming the branches and twigs at their intended victims. In captivity, great apes become very adept at hitting zoo patrons with objects, particularly feces. Using Beck's definition, the branches and foliage used in aggressive displays by male gorillas and chimpanzees could be classed as tools because they very effectively serve to alter the actions of conspecifics. The classic example of this type of tool use by a chimpanzee is the kerosene can banging display which was used by Mike to achieve the position of alpha male in the Gombe community (Goodall, 1985).

The Acquisition of Food

Chimpanzees, in particular, are noted for their creative use of tools to acquire inaccessible food items. Jane Goodall's observations of the chimpanzees at Gombe fashioning sticks as tools to fish for termites, made headlines in scientific publications and the non-academic press because it seemed that the uniqueness of human achievements was being questioned. Chimpanzees, in fact, use tools in a number of ways, although the different types of tool use seems to be specific to certain chimpanzee communities. While all of the East African chimpanzees at Gombe and Mahale Mountain fish for termites using sticks, most of the West African chimpanzee communities get at termites by perforating the nest and scooping them out with their hands. Only the Gombe chimpanzees use chewed-up leaves as sponges to sop up water from tree crevices or scoop brain tissue from the skulls of colobus monkeys. Ivory Coast chimpanzees use a hammer and anvil technique to get at the fruit of hard-shelled nuts. The nuts are collected and carried to particular areas where flat rocks or smooth tree roots can serve to position the nut so that it can be cracked with a heavy branch or stone (Tomasello, 1990).

Cebus monkeys have been compared to chimpanzees in their ability to use creativity in extractive foraging. Laboratory studies report a variety of intelligent use of tools, including raking in out-of-reach objects with sticks, rings and sacks, using cups to collect water, and prying open boxes with spoons. In the wild, their efforts are confined to cracking open nuts with rocks and fishing for ants and baby birds with their hands (Parker and Gibson, 1976).

Body Care

Cebus monkeys rub themselves with urine, presumably as a form of olfactory communication. However, they have also been observed anointing their bodies with a variety of materials such as ants, orange peel, and onions. The reason for this is not known, but the frenzied rubbing that follows the introduction of green onions into a cebus enclosure indicates that it provides some sort of stimulation, possibly to the skin. A number of other species, such as Japanese macaques, baboons chimpanzees and orangutans, use leaves to wipe away sticky and unpleasant substances from their bodies. The tendency of the great apes to adorn themselves with branches and leaves, or anything else that is available, is especially noticeable in captive situations. However, orangutans in the wild are noted for their purposeful use of broad leaves to protect them from the rain or hot sun.

Functions of Tool Use in Primates

Tool use in animals could also be viewed in terms of the function served instead of the context of the activity. Beck (1980) suggested that tools are generally used for one of four purposes: to extend an animal's reach, to magnify the force of an action, to augment displays, or for greater control of liquids. Most of the context-related examples of tool use previously mentioned fall into one or other of the categories. However, studies of captive primates have recorded extraordinary accounts of the use of ladders in escape attempts by chimpanzees (Menzel, 1972), and a variety of reach-extending devices by many primate species to achieve access to items other than food. An unusual leaf-clipping display that has been observed as a means of sexual solicitation in one of the chimpanzee communities at the Mahale Mountain site could also be viewed as a form of tool use. The display, in which a male takes several leaves and repeatedly rips them apart with his teeth, often serves to gain the sexual attention of females.

Parker and Gibson (1977) recognized social tool use as a way of achieving a goal by using an animal as a tool to change the behavior of another animal. A classic example of social tool use is the **agonistic buffering** observed in baboons, where an adult male will pick up and display a specific infant to defuse an agonistic encounter with another male. Hans Kummer (1982), suggested that social tool use requires more finesse than mechanical tool use, since the social tool has goals of its own and can change during the interaction.

The relationship between tool use and intelligence is difficult to assess, because, in some cases, the use of the tool does not seem to require much skill or knowledge. An even more vexing puzzle is why we do not see a greater use of tools in the wild, given the impressive creativity in the use of tools shown by primates in captive situations (Jolly, 1985). Jolly (1985:361) suggested that primates possess a general intelligence: "which allows one to combine more than one perception or idea, and thus create new concepts, which are revealed in new patterns of behavior." This intelligence can be directed towards practical purposes such as finding and accessing embedded food, or for social purposes, such as improving one's dominance rank or securing a mate.

The Evolution of Primate Intelligence

Primates are noted for their large brains, particularly in the neo-cortical area, compared to mammals of similar size (Jerison, 1973). Given that the brain is an extremely energy-expensive organ to operate, there must have been a major adaptive advantage for the expanded neo-cortical functions found in primates (Gibson, 1986), however there is some disagreement among primatologists as to the origin of this adaptive function. There are two main theories regarding the evolution of of the flexible, creative intelligence generally attributed to primates: the "social origin" theory and the "ecological origin" theory. Some researchers argue that ecological pressures, such as foraging strategies or the successful avoidance of predators, selected for higher intelligence (Gibson, 1986; Milton, 1988), while others suggest that the social pressures associated with group living provided the major impetus for the evolution of cognitive abilities (Jolly, 1966; Humphrey, 1976; Burton, 1984; Jolly, 1988). Since intelligence is a multi-faceted ability, likely to defy any theory that rests on only one factor in explaining its evolution, a case could be made for either point of view. Clearly, the relative contribution of ecological or social factors to the evolution of intelligence would vary with the species, depending on their type of social organization and their adaptive strategies.

Ecological Origin Theory

Milton (1988) reviewed evidence from a number of field studies to add validity to her argument that successful foraging behaviors provided the impetus for the evolution of primate intelligence. Her case rested on the following facts about food availability in the tropical forest:

a.   Diet is related to home range size. Since leaves are not as patchily distributed as ripe fruit, folivores generally require a smaller foraging area than frugivorous species.

b.   Potential plant food items are patchily distributed in space, with the great diversity of tree species distributed in clumps rather than in a uniform fashion.

c.   Potential plant foods are patchily distributed over time, with the availability of high quality food, such as ripe fruit, typically short-lived.

d.   The problems associated with the patchiness of the food supply in time and space are somewhat lessened because patchiness can be predicted, given a knowledge of the environment.

The last point may have acted as an important factor in the development of the complex primate brain, since the ability to predict the availability of plant food in the tropical forest likely requires a greater mental capacity than that required by carnivores who are secondary consumers of food. There would be selective pressure to promote the use of efficient strategies in the search for food, combining the lowest expenditure of time and energy with the least risk of exposure to predators. The need to respond to changing environmental conditions and to form contingency plans would also challenge mental capacities. This theory argues that species who are dependent on diverse, sparsely available foods should show greater mental capacities than those who rely on uniformly available resources. Field studies of howler and spider monkeys living sympatrically on Barro Colorado Island have been used to promote this point of view. The basic needs of the two species differ, with the howlers supplementing their folivorous diet with fruit in season, but reverting to a diet of leaves when food was scarce, whereas the spider monkeys depend on fruit regardless of the availability. The location of preferred food presents a greater problem for the spider monkeys and requires a more complex foraging strategy than that used by the howler monkeys, who seem to be able to exist on a wide variety of readily-available leaves. The advocates of this point of view refer to the fact that folivorous species, who presumably do not need sophisticated foraging strategies, generally have smaller brains relative to body size than frugivorous species. Milton (1988) noted that the close relationship between brain size and diet would be difficult to explain if social factors were responsible for expanded intelligence, especially since there is little relationship between brain size and sociability.

By virtue of their superior foraging skills, cebus monkeys provide further support to the ecologically-oriented theory. Parker and Gibson (1977) suggested that adult cebus monkeys demonstrate an ability for complex object manipulation that is not context-related and can be generalized to a number of situations. From the available data, it appears that this ability far exceeds that shown by any other nonhuman primate species, except chimpanzees. Cebus monkeys are able to recognize possible sources of embedded food and use a number of cortically-mediated coordinations to find, extract, and process food (Gibson, 1986). The monkeys appear to exhibit planning abilities in their foraging techniques. For example, they damage some fruits, which serves to speed the ripening process. The correlation between brain size (particularly the neo-cortex) and "omnivorous extractive foraging habits" (Gibson, 1986:97) is demonstrated by primate encephalization and neo-cortical indices. These measures were used by Stephan (1972) to refer to the extent that the size of the brain and the neo-cortex exceed that predicted for insectivores of similar size. In both of these indices, cebus monkeys are on a par with chimpanzees, ranking second only to humans (Stephan, 1972). However, the use of innovative foraging strategies and complex object manipulation does not appear to be related to omnivory alone, as macaque species, vervets and baboons, who are also highly omnivorous, do not appear to show the same degree of complexity in their manipulation of objects in natural situations. It is argued that inventive intelligence would be selected for in situations where

189

the availability of nutritious food was constrained by seasonal variations and inaccessibility (Parker and Gibson, 1977) (see Figure 15.1).

Figure 15.1   A white-faced capuchin (*Cebus capucinus*) eating some hard-won food item, closely watched by a conspecific. (Photo by Russell A. Mittermeier, Conservation International.)

Social Origin Theory

A similar argument for a relationship between relative neo-cortical size and social complexity can be used to promote the view that social pressures were important in the evolution of primate intelligence. Sawaguchi and Kudo (1990) found that there was a significant positive relationship between group size and neo-cortex size in prosimians. When they controlled for diet in anthropoid species, this relationship was also present, and the relative size of the neo-cortex was greater in frugivores living in polygynous groups than in monogamous species. From these findings, one could infer a close relationship between social structure and the degree of neo-cortical development, with the accompanying implication that larger, presumably more complex, social groupings could have selected for a greater mental capacity.

An increasing number of scholars from different disciplines have promoted the notion that social pressures were a potent force for the evolution of primate intelligence (Whiten and Byrne, 1988). In nonhuman primates, some of the most convincing evidence lies in the use of social knowledge to mediate interactions and to cement relationships. This volume has repeatedly emphasized the social nature of primates and the importance of maintaining social order, cohesiveness and continuity in their groups, in the face of changing physical and social conditions. Primate social relationships tend to be more complex than those in many other species, due to the extended period of dependency which allows for a greater degree of social learning than that available for shorter-lived species.

Harcourt (1988:111) asked "What are primates doing with their large brains and superior intelligence that non-primates are not?" He argued that primate physical environments and diets are no more complex than those of non-primate species, and superficially the primate social environment does not appear to be more complex. However, on closer inspection it could be noted that the types of social interactions observed in primates are more sophisticated, particularly those related to alliances and the three-way interactions frequently seen in primate societies. The sophistication of nonhuman primate social skills has been well-documented in the reports of field studies, particularly those species that have been studied on a long-term basis. Goodall's reports of the complexity of chimpanzee society are most convincing. Cheney and Seyfarth's (1980) innovative experiments on wild vervets, described in Chapter 14, have shown that these monkeys can readily recognize individuals in their social group and are aware of their own kin relationships as well of relationships between other members of the social group. Kin recognition, which is defined as preferential treatment towards known relatives, is a generally-accepted faculty of a number of taxa, including insects, reptiles and rodents. However, the recognition of the kinship relations of others has only been noted in primates (Gouzoules, 1984). Field work by Smuts (1985) and Strum (1987) clearly indicates that savannah baboons use sophisticated social strategies to improve their position in the group. Anestrous adult female baboons maintain lasting "friendships" with one or more males, providing added protection and care for themselves and their offspring. Male baboons, in particular, need extensive social skills to maintain their place in a given social group or to gain a place in another troop.

Dominance in primates is often discussed in terms of social attributes, rather than physical, attributes such as size and strength. There are numerous reports in the literature of old, ill, physically weak primates who were in a position of dominance in their social group. Fedigan (1992:102) argued that, although a strong healthy body would be an advantage, "The power of social learning and social tradition is so pervasive in primates that it often overrides the importance of physical factors in the determination of rank." Since a position of dominance often confers an advantage to the individual in terms of access to resources such as space, mates or food, the social intelligence required to achieve and maintain rank is important in many respects. Older high-ranking individuals not only appear to have social skills that allow them to influence others in the social group, but they often have knowledge of sources of water or food that could be sought out in times of drought. Hauser (1988) documented such a case in vervets, where a technique presumably to soften tough acacia pods was first observed in an old dominant female during a severe drought in Amboseli National Park. The technique, which involved soaking the pods in an exudate found in tree crevices, spread throughout the adult females and juveniles in the troop and was used extensively only during the period of water shortage.

From captive studies, it is apparent that nonhuman primates have mental capacities beyond those used in the wild, such as in the use of tools. It is possible that the function of this intellect could be socially-oriented, since living in a group, for all of its advantages, presents many problems—one of which is calculating the consequences of one's behavior in a social interaction. Humphrey, a philosopher who is one of the major advocates of the social origin theory, argued that animals living in complex social groups require creative intelligence to hold their societies together. Maintaining social life is one of the most urgent aspects in the organization of an individual's behavioral repertoire, since one must be able to assess others and make proper adjustments based on the behaviors of others (Humphrey, 1976). A number of primatologists have suggested that skill in managing social situations was a major force in the evolution of cognition and that "social life preceded and determined the nature of primate intelligence." (Jolly, 1966:506).

Humphrey's contention was that ecological pressures, even the need for inventive techniques for obtaining embedded food, could be managed by "practical" intelligence, and that techniques like termiting can be learned through observation and practice and do not require creative intelligence (Humphrey, 1976). It is possible that the practical knowledge required by primates can be readily acquired during socialization, when learning usually takes place.

Clearly, selection for the evolution of higher intelligence in primates was not an "either-or" proposition, and both ecological and social pressures were involved. As with every question addressing the evolution of behavioral attributes, we can only deal with speculations, and make educated guesses based on the available evidence from extant species. It is impossible to know the relative contribution of ecological or social factors to the evolution of intelligence in a given species. What can be said, however, is that intelligence in modern primates allows for a behavioral flexibility that can be applied in both social and non-social domains.

Social versus Non-Social Cognition

The discussion concerning the social competence of nonhuman primates could lead one to question the distinction between social and non-social cognition. Early developmental psychologists assumed that cognition was the result of the mental constructs that could be applied in any domain. Recently, however, primatologists and psychologists have suggested the possibility that there is a qualitative difference between social and non-social cognition. They point to the difficulties encountered in training laboratory monkeys to do tasks dealing with concepts like transitivity, while monkeys in the wild seem to learn their place in the dominance hierarchy (a similar concept) with ease (Cheney and Seyfarth, 1990). The experiments conducted by Cheney and Seyfarth on free-ranging vervets indicate that the same animals who are so adept at picking up social cues seem to be unable to use physical cues to alert them to the presence of a predator. This is not to suggest that the socially-skilled vervets are lacking abilities in all other areas, only that their intelligence may be further developed in the social arena.

Psychologists at the *Instituto di Psicologia* in Rome have conducted a number of detailed experiments comparing the cognitive development of primates. They tested macaques, cebus monkeys and gorillas in the development and use of the ability to deal with physical objects and compared the results to those reported for human infants. They found that cebus monkeys fall between macaques and human infants in terms of their manipulatory skills. Cebus infants showed more repetitive interactions with physical objects than the gorilla or macaque infants, but were less creative in their use of objects than human infants. Cebus infants showed consistently higher scores than macaques in object manipulation, and appeared to have a better grasp of cause and effect in their ability to retrieve items with a stick (Antinucci, 1989). Poti and Antinucci (1989) established criteria for first and second order cognitive constructions to test the logical abilities of their primate subjects. First order constructions involved the ability to put together a set of two or more objects based on some characteristic, such as color or shape. Second order constructions referred to the ability to construct sets of sets, where a set of objects is seen to be related to another set based on a common characteristic. Cebus monkeys were able to master second order constructions, while the macaque species never advanced beyond first order capabilities.

It would appear from the wealth of material attesting to the social attributes of macaques, baboons and vervets, that Old World species are quite capable of second order constructions in their social lives. The ease with which these primates successfully deal with their own kin, the kinship relations of others, and dominance hierarchies on a daily basis strongly indicates a well-developed ability to construct different sets of conspecifics based on social characteristics. Possibly the creative intelligence that is used so successfully by cebus monkeys in their dealings with environmental objects, is directed towards the social dimension in a number of other primate species, particularly those living in large, cohesive social groups. This leads to the speculation that the difference between social and non-social cognition, at least in nonhuman primates, is largely a function of the selective factors operating on the animals. Cebus monkeys, who live in dispersed social groups in tropical forests where food is sometimes difficult to find, have been selected to direct their creative intelligence towards non-social issues. Baboons, whose social sophistication is well documented, require extensive social skills to survive in their large, cohesive social groups.

Summary

Most primatologists agree that the mental capacities of nonhuman primates, particularly monkeys and apes, are impressive. Laboratory studies have provided ample evidence that many species are not only adept at several types of discrimination tests, but can solve puzzles, master abstract concepts and, in the case of the great apes, use gestural and artificial languages symbolically. Field studies record the extensive use of tools by chimpanzees, creativity in finding embedded food in cebus monkeys, and other examples of practical intelligence used on a daily basis by other primates. In some species, particularly those living in large, cohesive social groups like baboons, macaques species and vervets, the use of creative intelligence in social situations seems to surpass that used in a practical sense. Although it is tempting to speculate that social pressures may have been the major factor selecting for the creative intelligence and behavioral flexibility attributed to primates, an equally convincing case could be made for the view that environmental agencies provided the selective impetus.

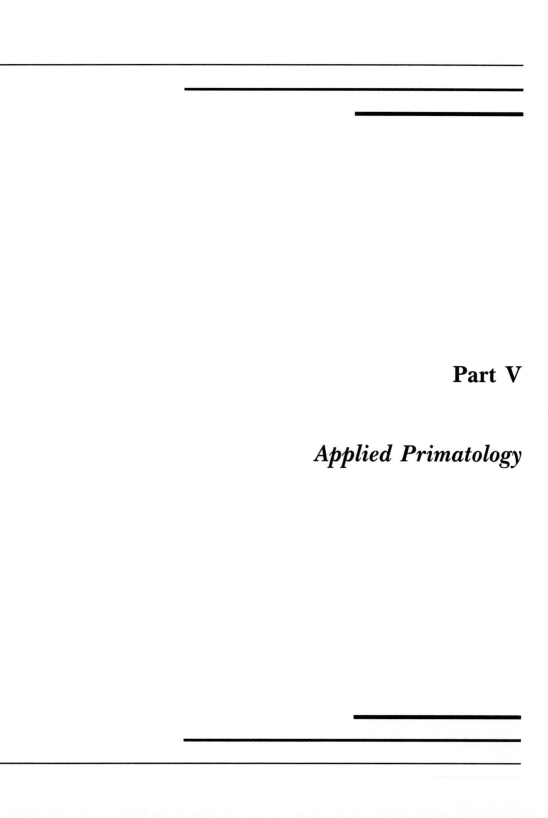

Part V

Applied Primatology

Chapter 16

Primate Ethology

Conducting a Field Study

Ethology is the study of animal behavior which is usually carried out under natural conditions. Much of the information contained in this volume is the result of ethological studies, involving untold hours spent in the field observing primates in their natural habitat. Video tapes of primates studies, like those of Jane Goodall and her chimpanzees, may present a glamorized version of what field work is really like. Field work can be very rewarding, but often these rewards are won by dints of hard work, physical discomfort, and not a little tedium. Ethology is a science, and if an ethological study is to be treated seriously, the primatologist must adhere to the guidelines that have been established for scientific research, which include: the description of the research question, isolation of the variables, data collection, data analysis, hypothesis testing and interpretation of the results (Lehner, 1979).

Setting up a Project

Generally speaking, there are two types of research projects conducted by ethologists:

1. Species-oriented research, in which the species is of particular interest and the objective is to find out as much about the primate form as possible; and

2. Concept-oriented research, in which the primary objective of the study is to investigate an aspect of behavior, such as agonism. In this type of a project, it is important to pick a subject species that best illustrates the researcher's particular interest. For example: Savannah baboons would be the species of choice over vervets for a study focusing on agonistic encounters becuse the latter species exhibits less overtly aggressive behavior.

Once the species has been established, and the relevant information on that species searched out, the next important decision that must be made is the research site. Basic to a successful study is the choice of a location that suits the researchers financial and physical needs. The accesability of the area where the animals are found, the stability of the government and its attitude towards researchers, the accessability of the primates in terms of habituation and visibility, and the climate are factors that must be taken into account before choosing a research site.

Having established the study site and reached the destination, the next problem facing the researcher is to find the study group or groups of primates and conduct reconaissance observations of their activity patterns, daily ranges and other relevent information. If the primates have never been studied before, the initial stage of locating the animals and habituating them to the presence of humans may be lengthy. A log book documenting these early observations and subsequent ad hoc field notes will be extremely useful to the researcher when the study is completed and the first impressions have begun to fade. The reconaissance observations are also critical in establishing an **ethogram**. The ethogram, which is a comprehensive catalogue of a species' behavioral repertoire, must be defined before data collection can begin. It is important that the behaviors in the ethogram are clearly described in terms of the motor patterns that make up the action, and not the observer's interpretation of the intent so that the study can be easily understood and replicated by other scientists. The behavioral units used in the ethogram depend on the focus of the study. In a species-oriented study, the researcher will try to include as many of the species' behaviors as possible, whereas in concept-oriented study, the ethogram is tailored to the behaviors that pertain directly to the research question. Ethograms include two types of behaviors, differentiated on the basis of their duration. Behavioral **states** are on-going behaviors that have a definite time-frame, such as foraging, and **events** are instantaneous occurrences with no measureable duration, such as a cuff or a bite (Lehner, 1979).

Methods of Data Collection

Sucessful data collection depends not only on the choice of a suitable sampling method, but also on the use of reliable equipment. The equipment can be as simple as a clip board, pencil, stop watch and binoculars or as sophisticated as a computerized recording device, camcorder and tape recorder. The choice depends on the the financial resources of the researcher and climatic conditions at the research site. Although video tapes are useful in observational studies because they provide permanent, verifiable record of the behaviors, the expense of taping months of research may be prohibitive. Information collected by means of computerized data recorders has the advantage of greater accuracy than hand-written data, but computers tend to break down in hot, humid environments. If the research site is in the tropics, one must decide whether to choose reliability over accuracy. Often, the simplest and most basic data collection devices are the most practical, especially in Third World countries, where batteries, video tapes and computer discs may not be readily available (see Figure 16.1).

There are a number of different sampling methods that can be used, depending on the research question. The following is a description of the data collection methods that meet most research needs (Lehner, 1979):

1. **Ad libitum sampling** - This method refers to the typical field notes in which the researcher record as much as possible by informal observation. This is commonly used in the initial stages of a study, and the researcher must keep in mind that these ad hoc observations can only be used for descriptive purposes and not for quantitative analysis.

Figure 16.1   Researcher taking data using only a clip-board and a stop watch.

2.   **Focal animal sampling** - In this method, one individual is the focus of the sampling session. All of the behaviors of the subject animal and the durations of the behaviors are recorded over a designated period of time. This is especially good for detecting the degree of behavioral variability between categories of animals such as different age groups, the sexes, and animals of different rank. The focal animal technique is frequently used because it provides more information than any of the other methods, providing the researcher with frequencies and durations of behaviors, as well as patterns and sequences of interactions. The main drawback is that it is very time-consuming, especially if a large number of animals are being sampled.

3.   **All occurrences** - This is the most efficient sampling method for the investigation of a few easily recognizable behavioral events, such as feeding activities, common to all of the members of a group. The occurrences of these behaviors are recorded during a specified period, providing the rates of behaviors in the group as a whole and information about the degree of behavioral synchrony.

4.   **One-zero sampling** or **Hanson's check list** -This one of the easiest sampling methods to use and the one that provides the greatest degree of **inter-observer reliability.** The method involves scoring whether or not a behavioral state or event occurs during a relatively short period of time (e.g., 15 seconds). Since the behavior is scored only once, even it occurs a number of times, this technique cannot be used to record frequencies or rates of behavior and is of little use in a quantitative analysis.

5.   **Instantaneous** or **scan sampling** - The observer records the behavior of one or more individuals at preselected intervals over a period of time (e.g., every 30 seconds for 15 minutes). This is useful for recording activities that are easily distinguished and is used for testing the degree of behavioral synchrony within a group and for estimating the percentage of time devoted to any one activity.

Once the data has been collected, it must be entered into a computer so that appropriate statistical tests can be carried out. It is beyond the scope of this volume to discuss the statistical analysis of the data. However, for a clear, concise description of ethological research, from the inception of the study to the data analysis and the interpretation of the results, the reader is directed to <u>The Handbook of Ethological Methods</u> by Phillip Lehner.

Although this section has been directed primarily to the organization of a field study, the methods of data collection discussed above are equally applicable to research conducted in the laboratory.

Laboratory Research versus Field Work

Nonhuman primates and other animals exhibit behaviors in a natural habitat that are never seen in captivity. This is not surprising because captive animals are usually assured of an environment, sheltered from the rigors of the weather, parasitic infestations and predator attacks, and do not have to forage for their food and water. This is not to say that useful behavioral studies can only be carried out on free-ranging animals under natural conditions. In the past, laboratory researchers and field workers were sometimes sharply divided in their views regarding the validity of each others' work. While field workers occcassionally insisted that the behavior of captive animals was aberrant and unnatural, those who worked in laboratories deplored the lack of control and scientific rigor in field studies. The latter argued that there was no such thing as a "natural habitat" because the mere fact that researchers were there indicated that the animals had come into contact with some degree of human disturbance. Fortunately, the schism between scientists has been largely put to rest, because it has become clear that captive and laboratory studies complement each other and provide us with much needed information about the life-way patterns of nonhuman primates. The problems associated with each type of research must be recognized and adapted accordingly.

Laboratory Research

Laboratory research can claim the following strengths:

1.   The main advantage of laboratory studies lies in the ability to control the environment and ensure the validity and reliability of the results. The behaviors are not affected by climatic conditions such as wind, rain or extreme heat. This factor also allows the researchers greater flexibility in the choice of equipment, because computers and other sophisticated instruments can be used effectively indoors.

2.   The variables can be easily manipulated, and different test situations established.

3.   The animals can be readily identified and their complete life histories are known, including relationships between the animals and their exact ages.

4.   The visibility of the animals is ensured, facilitating studies of intimate social relationships, such as the mother-infant bond.

5.   If social behavior is the focus of the study, the fact that the animals do not have to forage for their food allows them more time for social activities.

6.   Because the conditions in the laboratory can be manipulated and controlled, these studies are easily replicated by other scientists.

Despite these factors that ensure the scientific validity of the research, the behaviors of the study animals may be affected by the captive environment. It is sometimes difficult to say whether a particular behavior is part of the animals' normal repertoire or is a function of being kept in captivity. Clearly, laboratory animals do not show the full range of behaviors that are seen in the wild because they do not have to forage for their food, escape from predators, or adapt to changing climatic conditions.

Field Research

Field studies are particularly important at this time, given the fact that many nonhuman primate species are endangered or at least in some sort of jeopardy. We cannot possibly establish appropriate conservation measures or even breed them in captivity unless we know how they live in the wild. The main strength of field work is that we will be better able to understand adaptive patterns by observing primates in their natural habitat. Even though different habitats may affect the behaviors of the animals, it can be of great value to observe how the animals adjust to different environmental agencies.

The main problems associated with field work are related to the fact that most primates live in the tropical forest where observational research is particularly difficult. Locating groups of arboreal monkeys may be present the first problem. Once found, the monkeys may flee from the sight of an unfamiliar animal (the researcher) or the foliage may obstruct the observers view. Identification of individual animals may be difficult at first, and it may be impossible to be more specific about their age other than old adult, young adult, subadult and infant. However, despite the logistical problems, with the aid of marking devices, high-powered binoculars and a great deal of patience, most field workers have been able to conduct projects that adequately fulfill the requirements of scientific research. Long-term studies of the chimpanzees at Gombe and Mahale mountains, and of many troops of Japanese macaques have been able to circumvent many of the methodological problems and have provided us with a wealth of data on the ecology and social patterns of these species.

Summary

Our knowledge of the ecology, behavior patterns and the biology of nonhuman primates has increased enormously over the last decades due to the contributions of both laboratory and field researchers. Technological advances have improved the quality of data coming from the field, and efforts on the part of zoos to simulate the natural physical and social living conditions of their primates have contributed to a reduction in the disturbance and stress of captivity. When we begin to see similar behavior patterns in the reports of field and captive studies, then we can be sure that they are truly representative of a species repertoire.

Chapter 17

Primate Conservation

Over the centuries nonhuman primates have alternately fascinated and repulsed us. Now, even though they are firmly established in scientific circles as our closest living relatives, and public opinion is generally on the side of the monkeys and apes, we have a lot to answer for in respect to our treatment of these unique animals. We have not only used them for our own gratification, as stand-ins for humans in biomedical research and entertainment venues, but have allowed the destruction of vast tracts of tropical forest. If continued at the present rate, means certain extinction for many primates and all other animal species that depend on the rain forests. This final chapter will discuss the uses and abuses that we have inflicted on nonhuman primates as well as the issue of the conservation of primates and their environment.

Ethical Considerations - Uses and Abuses of Nonhuman Primates

It seems appropriate to begin this section with the following observation by Jane Goodall (1990:245):

> The more we learn of the true nature of non-human animals, especially those with complex brains and correspondingly complex social behaviours, the more ethical concerns are raised regarding their use in the service of man — whether this be in entertainment, as pets, for food, in research laboratories or any other use to which we subject them.

Traditionally, the uses to which we have put nonhuman primates and our attitudes towards them is reflected in the following six areas. They are: hunted (primarily for food); used as caricatures of ourselves in the entertainment industry; used as stand-ins in biomedical and psychological research; kept as pets; viewed as pests; and worshipped.

Hunting

The discovery of processed baboon bones in conjunction with those of early hominids in Africa indicates that the hunting of nonhuman primates has had a lengthy history dating back to prehistoric times. Although modern nonhuman primates have been hunted for a number of reasons, they are still pursued for food everywhere that they occur: in Africa, Asia and Central and South America. Thousands of primates are hunted yearly in most of the tropical areas of the world for food or for sale. In some areas of South America, monkeys provide over 25% of the meat in the diet of the natives. The effect of hunting for food on primate populations, however is not the same for all species. Since the size and palatability of the animals is a consideration, small species are hardly worth the effort and expense of the hunt. Hunting has devastated the populations of larger primates, such as the spider monkeys, in areas where there is still enough undisturbed forest for them to exist. The formerly sympatric howler monkeys are able to survive because of their unpalatable taste. Some species, such as the woolly spider monkey (muriqui), are in double jeopardy because the adults provide a tasty meal and the young are popular as pets (Mittermeier and Cheney, 1987).

In South-east Asia, pigtail macaques make up a significant part of the diet not only because they are good to eat, but because their meat is thought to have body-building powers (Bernstein, 1967). They are also in demand as part of the labor force to harvest coconuts, as they can be trained to climb trees and select and cut the ripe nuts. Some particularly adept monkeys can pick as many as 800 coconuts a day (Sitwell, 1988). The African black and white colobus monkeys are hunted not only for food but for their spectacular pelts which are used in ceremonial robes by the natives, and for tourist items. Although monkey teeth and skulls are often found in tourist shops in South America and Africa, these are often the by-products of hunting for food. Lowland gorillas are widely hunted for food and tourist souvenirs in West and Central Africa. However, the poaching of mountain gorillas for their hands and skulls has largely been curtailed due to the active tourist trade in live gorilla-watching (Mittermeier and Cheney, 1987).

Entertainment

The reports of the early Greeks and Romans being entertained by dancing monkeys indicates that the use of nonhuman primates as a source of amusement has been around for many centuries. Traveling minstrels in medieval Europe often kept trained animals, including monkeys, who would dance, juggle, tumble, walk on stilts and play musical instruments (Morris and Morris, 1966). Today, the animals of choice in the entertainment industry are chimpanzees, not only because they make ideal caricatures of ourselves, dressed up in clothing and acting like humans, but also because of their gregarious nature and their obvious pleasure in the limelight. The entertainers are young, appealing chimpanzees who appear to thrive on their trainers' attention and audience applause. However, this changes when they reach puberty. Adult apes tend to become aggressive and hard to handle, and often become so burdensome to their trainers that they are sold to biomedical laboratories (Linden, 1992).

The fate of these "over-the-hill" chimpanzee entertainers is only one of the ethical concerns voiced by advocates of animal rights, because the source of the young chimpanzees is often suspect. Despite bans on the sale of endangered species, the illegal trade in apes is still brisk. In order to capture an infant chimpanzee, the mother and likely several of the social group must be sacrificed, and then there is no guarantee that the infant will reach its destination alive. It is common knowledge that only about one out of ten young primates captured survives the mistreatment and stress of the journey.

204

Laboratory Research

Monkeys and apes have been of great value to mankind in biomedical research because of their physiological and genetic similarity to humans. Since the laboratory rat is on a different evolutionary line, it is more difficult to generalize the results of rat-based experiments to the human condition. The important role that nonhuman primates have played in the development of life-saving drugs, vaccines and surgical techniques cannot be underestimated. However, the future of the use of primates as laboratory animals is a concern, in view of the endangered status of many species.

In order to look towards the future, the history of the use of primates in biomedical research should be examined. In the 1950s and 1960s, hundreds of thousands of rhesus monkeys were exported from India to industrialized nations, particularly the United States, primarily for the development of a polio vaccine. In the 1970s, Indian conservationists noted that the populations of rhesus monkeys were declining at an alarming rate, and it became clear, due to the efforts of the International Primate Protection League, that the monkeys were being used for other than biomedical purposes. In 1978, the Indian Government halted the export of monkeys to the United States and from then on, it became necessary to either purchase other macaque species from Indonesia, or use monkeys from breeding colonies in order to continue research. Although some primates, such as squirrel monkeys, continued to be imported for experimental purposes, an international treaty banning the sale of endangered species greatly curtailed the trade in many species, and the main source of laboratory animals has become the breeding colony (Mack and Mittermeier, 1984).

Unfortunately, not all animal research is essential to medical progress, as some merely duplicates previous experiments, and other experiments are directed towards gaining interesting but non-essential knowledge. It is unrealistic to suggest that the use of animals in biomedical research should be discontinued entirely. However, a concerted effort can be made to cut down the number of animals sacrificed, or to use computer simulations or other methods of testing drugs and medical techniques. At the very least, the animals that are being used should be housed in suitable enclosures, with every effort being made to attend to their physical and social needs (Goodall, 1990).

There is encouraging evidence that medical research laboratories can become important facilities for breeding endangered species. The New England Regional Primate Center, for example, which is investigating the effects of diet on ulcerative colitis and colon cancer, has one of the largest captive groups of cotton top tamarins in the world. The research is relatively conservative and non-invasive, and the breeding program could become a major source of these highly endangered New World monkeys for zoos, or perhaps even introduction into the wild (Peterson, 1989).

Similar ethical concerns about the use of primates in biomedical research could be directed towards the ape language experiments. What happens to the pampered, signing apes when the research is discontinued or the funding runs out? Unfortunately, the answer often lies in laboratories for experimental medicine, as subjects for hepatitis or AIDS research, or in semi-forgotten cages in psychology laboratories (Linden, 1986). It appears that only a life-long commitment to their ape subjects, like that shown by researchers such as Francine Patterson and Roger Fouts, is the answer to this ethical dilemma.

Pets

Although nonhuman primates are cute and affectionate when they are young, they often become aggressive and difficult to manage when they reach puberty. This should not be surprising because primates,who are social animals adapted to life in a social group, find their natural inclinations thwarted by life in a captive situation, no matter how comfortable that life may be. The sharp canines, and strength and agility of adult monkeys and apes represent a real danger to family members. To compound the problems, young monkeys and apes are difficult to toilet train, and many species of prosimians and New World monkeys use scent marking as a means of communication wherever they happen to live. In addition to these aesthetic considerations, the close biological relationship between nonhuman primates and humans makes them susceptible to our diseases, and possible carriers of viruses, such as hepatitis, that can infect their human owners. Clearly, nonhuman primates make poor pets.

The pet trade in nonhuman primates has declined significantly since the 1960s, when import bans were imposed on user countries and export restrictions were instituted by source countries. The drafting of CITES (The Convention of Trade in Endangered Species of Wild Flora and Fauna) in 1973 has had a positive effect in curtailing trade in wild-caught primates. The countries that signed the convention agreed to ban commercial trade in the endangered species of plants and animals, and to monitor the state of potentially vulnerable species. Unfortunately, the effect of CITES has been weakened by a number of loopholes that allow countries to circumvent the convention. A major problem is that CITES is a treaty, not a law, which is administered through the member countries' own internal mechanisms, and some of the 87 member countries are more stringent than others in enforcing the bans. The fact that nations are able to exempt themselves from controls over a particular species has also been a cause for concern. Japan, Norway and the Soviet Union, for example, exempted themselves from controlling some whale products when they joined the treaty. However, despite its inadequacies, the CITES treaty has proved to be a very important document that has significantly reduced the trade of endangered species to developed countries (Peterson, 1989).

Combined with protective legislation in some Third World countries, it has also helped to control the present trade in threatened, but not endangered, primates. Unfortunately, it does not affect the local trade in developing countries, such as Brazil and India, who use primates for their own biomedical research and continue to have a large internal market for pets. In Brazil and Thailand, there is still a flourishing pet market dealing in species such as marmosets, squirrel monkeys, gibbons and macaques. Often, the live capture of young monkeys for pets is a by-product of hunting for food, which means that females with infants are the favorite targets for hunters. For slowly reproducing species, like the woolly spider monkey, this can mean almost certain extinction, unless some stringent measures are taken. In Brazil, a program aimed at educating the local people about the uniqueness of their primate resources has used the muriqui as its conservation symbol. With the continued support of the press, it is hoped that pressure from the people will help to suppress the pet trade, at least in Brazil. In Africa, the trade in young apes is particularly serious because it is is often hard for poverty-stricken local people to resist the huge prices offered by unscrupulous dealers (Mittermeier and Cheney, 1987).

Pests

The continued encroachment of human habitation and agriculture into tropical forested areas has resulted in a conflict between the local people and the nonhuman primates, whose habitat is being destroyed. The primates, being highly adaptable animals, often resort to crop raiding when their natural supply of food is curtailed. Although the effects of crop raiding are often overestimated by the farmers, they regard the monkeys as agricultural pest and do not hesitate to kill them. The most serious culprits are semi-terrestrial species like baboons and vervets in Africa and macaque species in Asia. These primates are opportunistic omnivores whose catholic tastes in food lead them to feed on crops and crop remains. Vervets, in particular, are able to survive in highly disturbed areas with a mosaic of forest and agricultural land, where other arboreal species have had to move out. These Old World monkeys, who were brought from Africa on slave ships, have managed to maintain large populations in the Caribbean islands of St. Kitts, Nevis and the Barbados, despite being heavily hunted as agricultural pests for several hundred years. The behavioral flexibility of primates is nowhere more obvious than in the survival techniques adopted by the normally vocal vervets, who have become silent and vigilant in agricultural areas (Fedigan and Fedigan, 1988).

As human populations increase and more tropical forest is converted to agricultural land, the clash between the primates and the farmers will likely escalate, which makes the task of conservationists in these areas much more difficult (Mittermeier and Cheney, 1987).

Worship

In some areas of the world, primarily East Asia, monkeys have been protected and worshipped for centuries. The earliest example of primate worship was in Egypt circa 1000 B.C. where male Hamadryas baboons were viewed as symbols of wisdom and virility. They appeared to lead pampered lives inside the temples and were mummified and buried with the priests, indicating their exalted position in Egyptian religious life. In present day India, the hanuman langur occupies an important place in the Hindu religion. The God Hanuman, who was a monkey, received the undying gratitude of the people by sending his monkey troops to save the kidnapped wife of the God Rama. The legend is still perpetuated, and the local people do not dare to harm the monkeys that are over-running their cities, however much they are tempted (Morris and Morris, 1966). A similar situation exists with regard to the status of the rhesus monkeys in India. The monkeys have an important place in the religious life of the Indian people, and although the people resent the presence of these noisy, aggressive animals in their temples and parks, they continue to leave food for the Gods, which in reality sustains the monkeys.

Clearly, the attitude towards nonhuman primates varies greatly in different parts of the world, Although those of us who live in developed counties generally do not condone hunting monkeys for food or shooting them as agricultural pests, we are not without guilt in our treatment of our closest living relatives. We tolerate the use of chimpanzees in circuses and television shows we use monkeys and apes for all sorts of interesting psychological research to add to our knowledge of human cognition, then relegate them to life in a biomedical laboratory, or worse. However, as Jolly (1985:459) remarked "These are sins of commission." The last section of this chapter is devoted to a discussion of some of our "sins of omission," in terms of our failure to protect the tropical forests, which can be even more devastating to primate populations.

The Tropical Forest

It seems paradoxical that, just as the interest in nonhuman primates is increasing (judging by the the coverage in scientific publications, the popular press and public television), the numbers of animals and species existing in the wild is decreasing at an alarming rate. Even a cursory glance at the conservation literature leads one to realize that, despite the substantial loss of primates through hunting and live capture, the major threat to primate populations is the destruction of their habitat. Over 90% of primate species depend on the tropical forests of Africa, Asia and Latin America for their survival. The disappearance of vast tracts of tropical forest inevitably means the disappearance of vast numbers of plant and animal species, including primates (Mittermeier and Cheney, 1987).

Tropical forests, which cover one billion hectares and make up 2/5 of the world's closed forest, are of primary importance globally because of their sheer size. Their role in the maintenance of world ecosystems and climate, although not totally understood, is known to be critical, and they are of world economic importance as a source of medicinal plants, plant foods and wood (Grainger, 1987). Despite the fact that tropical forests are vital to the ecology and the economy of the countries in which they occur, the tremendous growth in the populations of these same countries has created a situation where more and more land is needed to produce food. The population explosion in developing countries has contributed to a dependence on farming. Tens of thousands square miles of tropical forest are cleared each year for agricultural purposes, half of which will become useless scrub or grassland (Myers, 1984). The greatest threat to the tropical forests is the conversion of land to agriculture and ranching, which is, at best, a short-lived solution to the shortage of food. Farming in the tropics is very different from that in temperate zones, primarily because of differences in the quality of the soil. Tropical forests grow on very poor quality soils, containing few nutrients. Most of the necessary elements reside in the forest itself. Dead leaves and trees fall to the ground where they quickly decay and the nutrients are recycled back into the living forest. Traditional slash and burn agriculture works well in the tropical forest because small tracts are cleared by burning and the ash provides enough of the nutrients to sustain crops for a few years, after which the land is left fallow for a few years to allow it to regenerate. Temperate-zone agriculture, where the soil is sustainable, is not suitable in the tropics where the nutrient-poor soil is exhausted a few years after the the land is cleared for crops or for cattle ranching. Unfortunately, sophisticated farming techniques that could improve yields and perhaps sustain the soil are not available to farmers in undeveloped nations. Ranching operations rarely improve the lot of the local people as they often involve foreign-owned corporations who offer attractive economic packages to the local governments (Grainger, 1987).

The demand for fuel by the local people accounts for the removal of a significant amount of wood from the forests. Eighty percent of the two billion yards of wood burned for fuel each year is burned in the tropics primarily for cooking food. The problem lies in the fact that the demands for wood exceed the present supply, and this situation will likely to worsen. The tropical forests are also important sources of wood for paper products, as well as pulpwood and lumber. The export of tropical hardwoods has increased dramatically since World War II, bringing much needed money into Third World countries. Unfortunately, instead of carefully harvesting the trees on a sustained-yield basis, many countries have been plundering the forests with little thought for the future. Some of the forests are logged selectively, which should theoretically leave much of the forest intact. However, in actual practice, the heavy machinery used in selective logging often damages the remaining trees and opens up large areas of the forest for logging roads (Myers, 1984). In areas where the primate species depend on primary forest, selective logging is just as devastating as clear-cutting the forest (Mittermeier and Cheney, 1987). It has been predicted that the forests in Indonesia and West Africa will be depleted for commercial logging shortly and these areas will have to begin importing wood just to meet local demands.

One of the most damaging effects of logging, ecologically speaking, is the disruption of the water table. Without trees, the valuable top-soil is carried away during heavy tropical rains, ruining lowland crops, clogging reservoirs, and leaving behind dry, depleted land that is unfit for farming (Mittermeier and Cheney, 1987).

The depletion of the tropical forests not only affects the developing countries in which they occur, but has global implications, economically and environmentally. The forests represent an important genetic repository for medicinal and food plants. Many of our cultivated food plants, which originated from tropical forest species, still depend on the wild varieties to widen the genetic base for the development of disease-resistant strains. A number of commercially-important fruits are still only available in the tropics. The tropical forest is also the source of many of the important drugs that are in use today. Toxins, in the form of alkaloids produced by many tropical plants as a protection against predators, have been found to be potent agents for the treatment of human diseases and health problems. Quinine, for the treatment of malaria and rawdixin used to treat hypertension and stress are only two examples of medically important alkaloids. Although pharmacuetical firms have been able to produce some of these drugs synthetically, many of active principles are so complex that only extracts from the tropical plants can be used. The search for new drugs to fight cancer, heart disease and other killers continues, and if the tropical forests disappear, we will never know what life-saving medicines have been lost to us (Humphreys, 1982).

The global environment is closely tied to the tropical forests, as they act as vast ecological barriers to modify not only the local effects of sun, wind and rain, but also provide a buffer to absorb some of the carbon dioxide that is destroying the ozone layer. The **ozone layer,** which is a blanket of gases surrounding the planet, insulates us from the suns rays and provides a hospitable climate for plants and animals. Over the centuries, human activity has increased the amount of atmospheric carbon dioxide, changing the composition of the earth's atmosphere and reducing the insulating effect of the ozone blanket. The ultimate result of this action is global warming, or "the greenhouse effect," which will eventually have a devastating effect on world rainfall patterns and sea levels (Myers, 1984).

Conservation Priorities

The **Red Data Book** published by the I.U.C.N. (The International Union for the Conservation of Nature and Natural Resources), which is the major reference for endangered plants and animals, lists three categories of threatened species: endangered, vulnerable and rare. Although over 50% of primate species are threatened to some degree, the following section will only discuss the species that are in the greatest jeopardy and the regions that have the highest priority status in terms of conservation measures.

Central and South America

The most critically endangered species of New World monkeys are the wooly spider monkey, the only member of the genus *Brachyteles,* and all three species of the golden lion tamarin (genus *Leontopithecus*). Both are found only in small pockets of forest in the Atlantic region of Eastern Brazil, which is also one of the highest conservation priority areas in South America, and perhaps the world (Mittermeier and Cheney, 1987). The Atlantic tropical forest area, which has been reduced to a small fraction of its original size,is home to 21 species and subspecies of primates, 15 of which are endangered (Santos et al., 1987). Unfortunately, some of these species do not occur in the wild anywhere else. The forests are still being logged out or modified for cocoa production, with little government control or concern for the future (Mittermeier et al., 1987).

The woolly spider monkey, which is the largest New World monkey, is popularly called the muriqui and has become the symbol of the conservation movement in Brazil (see Figure 17.1). It has been estimated that in the year 1500, when Brazil was discovered, 400,000 muriqui ranged continuously throughout the Atlantic forests of Eastern Brazil from the southern state of Bahia to the coastal mountains of Sao Paulo. By 1972, the population had declined to an estimated 2,000. Although the monkeys have been heavily hunted over the centuries for food and pets, the major reason for the dramatic decline in numbers has been the loss of habitat. Where smaller, more adaptable species of New World monkeys, like marmosets and capuchins, are able to adjust to a disturbed habitat, the slower breeding muriqui is unable to live in fragments of forests (Mittermeier et al., 1987). Perhaps a more disturbing note is that these large monkeys are difficult to keep in captivity and the possibility of establishing breeding colonies to bolster wild populations seems remote.

210

Figure 17.1   Woolly spider monkey or miriqui (*Brachyteles arachnoides*).

The golden lion tamarin, which is also endemic to the southern state of Bahia, in Brazil, is now restricted to small areas of forest and a few poorly protected reserves (see Figure 17.2). The situation has been made even more serious through the decimation of wild populations by the illegal trade in these popular little monkeys. In 1983–1984, almost half of the total population was illegally sold to dealers outside of South America. Fortunately, due to the efforts of groups interested in the breeding of endangered species,

Figure 17.2   Golden lion tamarin (*Leontopithecus rosalia*). (Photo by Brian Keating. Courtesy of the Calgary Zoological Society.)

211

many of these animals were returned to selected breeding institutions in United States and South America. Perhaps the best hope for the future of the golden lion tamarin are responsible breeding programs, and the possibility that some of the animals can be successfully introduced into protected forest areas of Brazil (Mallinson, 1987). In the meantime, conservationists are attempting to educate the local people, particularly the children, about their indigenous flora and fauna, and the importance of preserving them for future generations.

Other very endangered genera of New World monkeys are the *Ateles* (spider monkey) and *Lagothrix* (woolly monkeys), who range widely and require large areas of undisturbed forest to sustain them. They are also heavily hunted for food, and their relatively large size and slow rates of reproduction make them particularly vulnerable to any kind of exploitation. As a result, they have disappeared from areas of forest that would normally sustain them (Mittermeier and Cheney, 1987).

Africa

Of the 55 species of nonhuman primates recognized in Africa, 14, are in some jeopardy according to the Red Data Book. This figure includes: all of the great apes, three subspecies of gorillas, the the common chimpanzee and the pygmy chimpanzee (Mittermeier and Cheney, 1987). Although Africa contains a smaller area of lowland rain forest than dows South America or Asia, it supports a greater diversity of primates than either continent. In some areas, such as the Kibale forest, seven to ten species of monkeys, one or two ape species, and two to five prosimians live sympatrically. As in other developing countries, it is the rain forest areas that are under the greatest pressure from human populations for agricultural land and economic development. Records show that while the human population growth in Africa from 1980 to 1990 was greater than in tropical South America and Asia, the level of food production was lower. Since the people in many African countries suffer from persistent poverty and hunger, the primates are under pressure from hunting for food in areas where the forest is still relatively intact (Oates, 1985).

The most endangered African species is the mountain gorilla, whose distribution is limited to two small pockets of forest in the Virunga mountains (shared by Uganda, Rwanda and Zaire) and the Bwindi forest (Uganda). The total population is estimated to be under 450 animals, which is extremely precarious given the fact that they reproduce so slowly—a female may only have two or three surviving infants over a lifetime (see Figure 17.3). Gorillas who feed heavily on ground vegetation and prefer areas of primary and secondary forest, compete with the local people for space. Although the mountain gorilla is protected in reserved areas throughout its range, the continuing growth in human populations makes the protection of these parks very tenuous. Rwanda is the most densely-populated country in Africa and although the *Parc National de Volcans* has been set aside as a reserve for the gorillas, the government has been threatening to take away some of the park for much-needed agricultural land (Peterson, 1989).

Since all of the captive animals are lowland gorillas (see Figure 17.4), the survival of the mountain gorilla is in the hands of the local governments who have the power to enforce anti-poaching laws and keep the reserves intact. The Mountain Gorilla Project,

212

Figure 17.3 Infant mountain gorilla (*Gorilla gorilla beringei*). (Photo by Brian Keating. Courtesy of the Calgary Zoological Society.)

funded by international conservation agencies, has provided a ray of hope for the mountain gorilla by habituating a few groups of gorillas to humans and creating a carefully controlled gorilla-viewing tourist attraction. The influx of tourist dollars has improved the Rwandan economy dramatically and may persuade the government to re-evaluate their plans to encroach on the *Parc de Volcans* (Peterson, 1989).

Although there are a number of semi-terrestrial, open-country species of primates in Africa, the more diverse tropical forest forms present the greatest conservation problem

Figure 17.4 Lowland gorilla. (Courtesy Linda Fedigan.)

due to loss of their forest habitat. The forests of West Africa, particularly Camaroon and Equatorial Guinea, represent conservation priorities because of the wide diversity of indigenous primates and the extent of the destruction of the forest. African countries, like Zaire, where the primates are still abundant and forests are relatively undisturbed, must be the focus of immediate conservation measures if the status quo is to be preserved (Mittermeier and Cheney, 1987).

Madagascar

All of the 20 to 25 species of lemurs living in Madagascar today are in some sort of jeopardy, making this area the highest conservation priority in the world. The forested areas of Madagascar sustain a higher diversity of primates than forests of similar size elsewhere, partly because the lemurs have filled the niches of other vertebrates that are not represented on the island. Since humans arrived in Madagascar about 1500 years ago, 14 species of lemurs have become extinct. Although over-hunting may have precipitated the demise of these species, widespread habitat destruction is now the primary threat to the primates. The forests have been cut down for a number of reasons, including wood for fuel and land for agriculture and cattle, all of which are aimed at fulfilling the basic needs of the people, rather than economic greed (Richard, 1982).

The aye-aye (*Daubentonia madagascariensis*), the only living member of its family, was once widely distributed in Eastern Madagascar, but is now so close to extinction that population estimates are not even available (see Figure 17.5). The strange appearance and crop-raiding propensities of this squirrel-like primate have had as much to do with its threatened status as the destruction of its habitat. Even now, although the aye-aye is protected by law, local people continue to hunt the few remaining animals because they regard them as evil omens or as agricultural pests. Nose Mangabe, a small, uninhabited island, has been set aside and planted with mangos and coconuts in an effort to sustain the aye-aye population. Hopefully, the few animals in captivity at Duke Primate Center will be breed as a buffer against extinction in the wild (Peterson, 1989).

Figure 17.5 Aye-aye (*Daubentonia madagascariensis*). (Courtesy David Haring.)

Asia

Of the 50 to 55 primate forms recognized in Asia, approximately 30% are threatened in some way. These conservation priorities include the orangutan, the only great ape endemic to Asia, four species of gibbons, nine monkeys and two types of tarsiers. The Asian primates presents a somewhat different set of conservation problems than those found in other areas because they inhabit a variety of arid and temperate zones, and a number of islands with low, isolated populations, as well as tropical forested areas. There is no other continent in which the nonhuman primates have coexisted in close association with humans for centuries. Attitudes of the local people vary from regarding them as objects of worship, as in the cases of rhesus macaques and langurs of India, to viewing them as a much-needed source of meat or money. However different the situation of the Asian primates may be, the basic conservation problems are acute for the same reasons; the extensive habitat destruction for the logging industry and the needs of the expanding human populations (Eudy, 1987).

Until recently, the lion-tailed macaque (*Macaca silenus*), which is the only truly arboreal species of macaque, was very close to extinction in its native India, due to large-scale logging and heavy hunting for food. Although the Indian government prohibited the hunting of the macaques in the 1970s, the laws were were largely ignored and the exploitation continued. Fortunately, they breed well in captivity and worldwide populations have increased from 400 animals in 1972 to 2,000 in 1982, due to carefully controlled captive breeding programs (Wolfheim, 1983).

All of the members of the genus *Rhinopithecus, the snub-nosed monkeys*, are perhaps the primary conservation priority in Asia at the present time. According to the Primate Specialist Group of the I.U.C.N., the three species, which are endemic to China, have the highest priority rating of all Asian primates. The future of *Rhinopithecus roxellana*, the golden snub-nosed monkey, is relatively bright because of its fecundity and the fact that it often shares the range of the protected giant panda. The other two species, whose populations are 500 or less animals, have smaller ranges and are not as well protected as the golden monkey (Eudy, 1987).

Indonesia has the greatest diversity of Asian primate species in the most problematic area, because of the huge human population and the extensive commercial logging operations that have been exporting tropical hardwood for decades. Most of the lowland forest is gone and the plight of the primates species is severe. The orangutan, the only truly arboreal great ape, is in jeopardy because its distribution is now restricted to rapidly disappearing areas of tropical forest on the islands of Borneo and Sumatra. Unfortunately, there is a movement to begin logging the few nature reserves that have been set aside for the orangutans. The two projects that have been set up for the rehabilitation of captive orangutans at Sepilok and Tanjung Puting in Borneo have been successful in that they are popular attractions for the Malaysans and have increased the local awareness of the plight of the apes. They have been less successful in their programs to reintroduce the animals into the wild because of the limited areas of forest available. Within Indonesia, the Mentawai Islands, which have the the highest number of endemic primates per unit area of all islands worldwide, must also be considered a top conservation priority due to extensive logging and hunting (Eudy, 1987).

Present Conservation Measures and Future Prospects

The major problem facing conservationists is to find a way of balancing international concerns about the preservation of primates and the tropical forests with the needs of the nations in which they occur. However much we need the tropical forests as sources of wood, food products, medicinal plants and the primates that we admire, it is difficult to tell the local people to stop clearing the land for the crops and wood that they desperately need. Clearly, our concerns are well-founded, since vast tracts of magnificent forests are being cleared for the sake of a year or two of farming or grazing. Thus, the international community must take the initiative and provide money and support for conservation programs. Before anything can be done in terms of conservation projects, our scientists and educators must help the Third World countries to understand their ecosystems and the problems associated with the clearing of the rain forests, so that some sort of land planning can take place. In order to sustain the land that has been cleared, cropping systems that are appropriate to rain forests must be researched and developed. In order to protect the forests that still remain, work must be done to institute programs for managing and regenerating the forests (Grainger, 1987).

Although the future for primates and their habitats does not look too encouraging, it is far from hopeless. Dedicated primatologists have been developing conservation-oriented plans, some of which have already been put into action. The Primate Specialist Group of the I.U.C.N. has prepared action plans for the conservation of the primates and their habitats in the continents where they occur. These plans include cataloguing the number of primate forms represented in each area and defining their patterns of distribution. The major goal of the plan is to maintain the diversity of primate species by ensuring the survival of endangered species and by protecting the large numbers of primates in areas of high diversity. Although it is clear that substantial areas of tropical forest will continue to disappear, the Primate Specialist Group hopes to curtail the loss of primates and their habitats by:

1. Setting aside protected areas for endangered and vulnerable species;

2. Creating large national parks and reserves in areas of high primate diversity and/or abundance;

3. Maintaining or increasing the effectiveness of parks and reserves that already exist; and

4. Creating and increasing public awareness of the need for primate conservation and the importance of primates both as a part of the natural heritage of the countries in which they occur, and as important components in environmental systems whose proper functioning is vital to human well-being (Oates, 1985; Eudy, 1987).

Although it is clear that the establishment and maintenance of protected areas for primates is their best chance for survival, only a small proportion of the world's primate populations live in parks or reserves. Therefore, the fourth point perhaps ranks as the first

216

priority, because if the local people and their governments do not understand that the needs of the human and nonhuman primates are not all that divergent, then there is little hope of attaining the other three objectives (Mittermeier and Cheney, 1987).

Russell Mittermeier deserves special mention as a primatologist who has devoted his life, since graduation from university, to the conservation of endangered species. In his capacity as a director of the Primate Division of the World Wildlife Fund, he has made regular visits to the highest priority areas in terms of primate conservation pressures (Peterson, 1989). His enthusiasm and skill in dealing with Third World governments has paved the way for the establishment of many of the conservation projects in progress today. Mittermeier is particularly concerned with the problems of hunting and habitat destruction that have decimated the populations of a number of New World species, and has worked closely with Adelmar Coimbra-Filho in the development of the primate conversation efforts in Brazil. Primate Conservation, the newsletter and journal of the Primate Specialist Group of the I.U.C.N., which is chaired by Mittermeier, provides current information concerning the status of primate populations and the progress of worldwide conservation projects.

The previously-mentioned efforts of the Mountain Gorilla Foundation have managed to make the local people aware of the mountain gorilla and its desperate situation through an intensive education program. They have also convinced the Rwandan Government to shelve their plans to take over some of the *Parc de Volcans* for cattle grazing, given the lucrative tourist trade in live gorilla-watching. The international attention directed to Dian Fossey's all-out efforts on behalf of her beloved apes, combined with the enterprising activities of The Mountain Gorilla Foundation, may be the salvation of at least one primate species (Peterson, 1987).

An example of a successful project engineered by a primatologist, with the help of the local government and a number funding agencies, was the opening of the Ranomofana National Park in southeast Madagascar in May 1991. Patricia Wright, who has researched primate behavior and ecology in South America and Asia, as well as Madagascar, was asked by the Malagasay Government to assist in the establishment of a park. A major factor in the choice of a suitable location for the park was the diversity of species that could be protected there. At least twelve species of lemurs, including the severely threatened aye-aye, are known to live in this particular area of southeast Madagascar. The master plan, which had to integrate the economic needs of the local people with the strategies for the protection of the animals and the tropical forest, included:

1. Sustainable alternatives to the slash and burn agriculture practiced by the people living on the periphery of the park. These included the planting of gardens and fast-growing endemic trees for fruit, firewood, building needs and shade, and the institution of rice paddies;

2. A survey of the the diversity of plants and animals within the park;

3. Forestry research;

4. Education of the local people with respect to the reasons why the forests and the animals must be preserved;

5. Assessment of the health needs of the local people in terms of nutrition and disease; and

6. The building of facilities to encourage ecotourism which would generate income to help the park to pay for itself.

It is hoped that the steps followed in the establishment of this park, which integrated conservation priorities with the priorities of the local people, will provide a blueprint for the foundation of future parks and wildlife reserves (Wright, 1992).

Jane Goodall has become a disciple for the protection of chimpanzees. She spends much of the year traveling the world, lecturing about the plight of the chimpanzee and raising funds for her Jane Goodall Institutes which are involved in a number of conservation projects in Africa. These projects emphasize the importance of involving the local people in their conservation activities. In Tanzania, natives are being used as "Chimp Guards" to monitor the movement of the chimpanzee groups near the local villages, and, with proper training, will be able to collect data on chimpanzee behavior. Some groups of chimpanzees are being habituated to the presence of humans, so that the ecotourism, that has been so successful in Rwanda can be tried in Tanzania. The Jane Goodall Institute has been instrumental in setting up sanctuaries for ex-pets and confiscated chimpanzees wherever possible (Goodall, 1990).

From this very brief overview of successful conservation projects that have been undertaken, it is clear that the efforts of primatologists are making a difference in the preservation of the primates and their habitats. It is perhaps the field researchers working in tropical areas, who incorporate a conservation component into their study and enlist the support of the local governments in increasing public awareness about the importance of the primates and the tropical forests, who will ultimately be our best hope for the future.

Although conservation projects in the tropical areas where the primates occur are critical, zoos in developed countries can also become important facilities for breeding endangered animals and for educating the public about conservation. It is the youth that will ultimately be saddled with global conservation problems, and where better to teach them the need for protection of threatened species than in zoos where they can come face to face with the animals involved. Captive breeding programs in zoos not only help them maintain their own populations, but provide important information about the reproductive biology of different primate species. Zoos now realize that breeding animals in captivity is a complex operation, and that it is not enough to simply put a male and a female together in an enclosure and expect them to mate. In order for mating to take place, the climatic, spatial, nutritional and social needs of the animals must be taken into consideration (Kleiman, 1980). It is to be hoped that the captive breeding of endangered species will provide animals that can be successfully introduced into their native habitats, provided some of these areas still exist. Adelmar Coimbra-Filho, a pioneer in the field of primatology in Brazil, and Devra Kleiman of the National Zoo in Washington attempted this type of program with the golden lion tamarin. However, few of the animals introduced into a protected area in Brazil have survived. The establishment of successful breeding colonies of golden lion tamarins and other endangered New World species in Brazil, particularly in the Rio de Janiero Primate Center, has been largely due to the efforts of Coimbra-Filho (Peterson, 1989). Now, several zoos in United states are involved in breeding the endangered tamarins. The collection of reproductive data, and the maintenance of an updated stud book will be important contibutions to the conservation of these attractive primates (Mallinson, 1987).

Zoos can be important sources of information on the genetic bases of the captive populations of the endangered species. The total worldwide captive populations of some primate species, such as the lion-tailed macaque, are derived from two or three founding animals, and although some zoos have large groups of these highly endangered monkeys, the genetic basis for the species is extremely narrow and they have become severely inbred. The American Association of Zoos has developed a computerized data base of zoo animals that lists the number of animals, their ages and sex of every species kept by each member zoo. This has proven to be a useful program for the arrangement of breeding loans and the exchange of information. This association has also set up Special Breeding Plans for threatened species, such as lion-tailed macaques, several species of lemurs and the western lowland gorilla, which involve setting up reliable genealogies of the captive animals and determining the carrying capacities of zoos.

Duke University Primate Center in North Carolina participates in the Special Breeding plans for some species of lemurs and has the largest breeding colony of prosimians in the world. Twenty species, representing 12 genera, are housed in various types of habitats, ranging from small nest boxes to large forested areas. The special care that is taken to provide the proper climate and spatial and social needs of the prosimians has been rewarded by the production of about 110 infants yearly (Anderson and Brown, 1984). Unfortunately, the two most endangered lemurs, the indri and the aye-aye, are difficult to raise in captivity and the likelihood of developing captive breeding programs is slight.

Zoo conditions for captive orangutans and lowland gorillas have improved greatly over the years. In the past, many of the captive apes who were were housed in bare, cold, featureless enclosures either died or became obese, sluggish animals. Modern zoos who have paid special attention to the needs of these animals have had success in breeding both of these apes. Orangutans are arboreal apes who require trees or some sort of climbing apparatus to keep them occupied, while gorillas are best suited to a grassy, leafy habitat with some access to the outdoors. The orangutans are now breeding well and are among the best represented endangered species in captivity (see Figure 17.6). The captive gorilla population is increasing due to the efforts of institutions like the Lincoln Park Zoo in Chicago, and since the Species Survival Plan has been instituted for the western lowland gorilla, the captive populations will be carefully managed and controlled (Peterson, 1989).

Concluding Remarks

Although the statistics reported for the rate of destruction of the tropical forests and the declining populations of primates seem grim, the conservation efforts of agencies such as the World Wildlife Fund, Conservation International, The I.U.C.N., The International Primate Protection League and The American Association of Zoos, are beginning to show

Figure 17.6   Female Sumatran orangutan and her infant. (Photo by Brian Keating. Courtesy of the Calgary Zoological Society.)

results. Much of the credit for conservation programs directed towards the protection of primates must be given to the many dedicated primatologists working in the field and in laboratories all over the world. These men and women have not only supplied us with the necessary knowledge about primate physiology, behavior and ecology, but have worked with local governments, instituted education programs and increased local as well as international awareness about the need to preserve the primates and their habitats. This volume is dedicated to these scientists.

Adaptive radiation - The rapid formation of a number of species with different adaptive patterns that occurs when a population or a species enter a new environment in which there is little or no competition.

Age-graded society - A primate society made up of one breeding male, his subadult sons, several females and infants, e.g., Gorillas society.

Agonism - Conflict behavior directed at conspecifics that is either self-assertive of self-defensive.

Agonistic buffering - A male picks up an infant and uses it as a sheild to defuse aggression from another male. Commonly seen in common baboons.

Allomothering - A female other than the mother cares for a young animal.

Allonursing - A female, other than the mother, allows an infant to suckle.

Alloparenting - A male or female caretakes an infant that is not his or her own.

Alpha - The highest ranking male or female in the social group

Altricial - Helpless at birth.

Altruistic - A behavior that helps the recipient while putting the donor at risk.

Analogical reasoning - The ability to understand relationships between physical or social objects.

Androgens - Male hormones

Anthropomorphism - Using human motivations and emotions to describe animal behaviors

Aphasic - Unable to speak.

Arboreal - Tree-dwelling.

Archaic primates - The first primate-like mammals that appeared on the earth during the Paleocene Epoch (65–53 mya).

Bicornate Uterus - A uterus in which the two Mullerian ducts are unfused. Seen in tarsiers and prosimians but not in anthropoids.

Binocular vision - The ability to see simultaneously with both eyes.

Bipedalism - The ability to locomote in an upright poition on two legs.

Biome - A geographical area that is characterized by a broadly similar type of vegetation.

Brachiation - A mode of locomotion characterized by an arm over arm progression through the trees.

Carotid Artery - The large blood vessel that carries blood to the brain.

Catarrhini - The Infraorder of primates that includes Old World monkeys, apes and humans. The term refers to the downward directed nostrils characteristic of the members of the Infraorder.

Cognition - Intelligence.

Cognitive Ethology - The study of animal intelligence.

Complete Competitors - Species that utilize the exact ecological niche, in terms of time, space and the use of resources.

Consort Bond - An intense temporary bond between a male and a female of the same species, usually based on a sexual attraction.

Conspecific - A member of the same species.

Continental Drift - The movement of the earth's crust over the partially molten interior which has caused the continents to change location over time.

Core Area - The portion of a species' home range that is used most extensively. It is usually the area in which food and safe sleeping sites are most available.

Day Range - The area over which a group of animals normally travels over a period of a day.

Dental Comb - The incisors and canines in the lower jaw which project forward to form a scraping device used for grooming and foraging. Found only in prosimians.

Dental Formula - This indicates the number of teeth of each type, listed from left to right (incisors, canines, premolars and molars) in one side of the mouth only.

Dietary Plasticity - The ability to survive on a number of different types of food.

Diurnal - Active in the daytime.

Dyadic - An interaction between two individuals.

Ecology - The study of an individual's relationship with the environment

Endocrine - Glands that produce hormones and secrete them directly into the bloodstream.

Epigamic Selection (Intersexual Selection)- The principle in the Theory of Sexual Selection that refers to interactions between the sexes. Usually discussed in terms of female choice for mates.

Estrus (noun) - The hormonal state of a female who is motivated and willing to mate.

Estrous (adjective) - Refers to the state of estrus.

Ethogram - A comprehensive cataolgue of a species' behavioral reperroire.

Ethology - The study of animal behavior. An ethological study is usually carried out in the field.

Exclusivity - A females' reluctance to allow her infant to be handled or looked after by another individual.

Facultative - Used occasionaly.

Fecundity - Refers to the ability of a female to produce infants.

Focal animal sampling method - Data collection method in which everything that one particular animal does over a specific period of time is recorded.

Folivore - Leaf-eater

Frugivore - Friut-eater.

Genotype - The genetic makeup of an individual.

Haplorhini - Dry nosed primate (tarsiers, monkeys and apes).

Heterodent - Having various types of teeth that serve different functions

Herbivore - An animal that eats only vegetable matter.

Home range - The total area utilized by a primate group over an extended period of time.

Hyoid bone - Bone that lies between the chin and the thyroid cartilage. Enlarged in howler monkeys to produce a capacious resonating chamber.

Inclusive fitness - The ability of an organism to reproduce and contribute its genetic material directly to the next generation plus the ability of its biological relations to produce offspring (an indirect contribution of the organism's genes).

Individual fitness - The ability of an organism to produce offspring and have its genetic material represented in the next generation.

Innate - The biological potential to behave in a certain way depending on environmental influences.

Instinct - Biologically determined behavior which is unaffected by environmental influences.

Inter-observer reliability - The degree of consistancy between the data taken by different observers in the same study.

Intrasexual selection - The principle in the Theory of Sexual Selection that pertains to interactions between members of the same sex. Usually discussed in terms of male-male competition.

Ischial callosities - Callous-like pads on the hind-quarters of Old World monkeys and lesser apes.

Kin selection theory - Individuals will act altruistically to a biological relation to assure that the portion of their genetic material shared with by kin is passed on to the next generation.

Mandibular - Pertaining to the lower jaw.

Matrilineage - A group of individuals who are related by virtue of common descent from a female ancestor.

Matrifocal Unit - A group consisting of the mother and her offspring.

Monogamy - A mating pattern in which a male and a female mate exclusively with each other.

Monomorphism - The males and females of a species have the same physical and behavioral characteristics e.g. in size, dental morphology and color.

Morphology - Physical form.

Multiparous - A female who has had more than one infant.

Natural selection - A mechansim for evolutionary change whereby the genes for the characteristics best suited to life in the environment are carried over to succeeding generations.

Neo-natal coat - The distinctive coat of an infant, which lasts for approximately the first ten weeks of life.

Niche - The environment exploited by an organism

Nocturnal - Active at night.

Nuchal crest - A ridge of bone in the occipital areaa of the skull that results from the attachment of the neck muscles.

Nulliparous - a female who has not had an infant.

Olfactory Sense- The sense of smell.

Ontogenetic - Developmental

Omnivore - Individual that eats a variety of plant and animal foods.

Ozone layer - A blanket of gases that surround the earth insulating us from the suns rays and providing a hospitable climate for plants and animals.

Parental certainty - The situation in monogamous species, where the male is sure that he is the father of the offspring.

Parent-infant conflict theory - The nature and extent of parental behavior can be predicted by viewing them as a function of the costs to the parents in terms of producing future infants and the benfits to the survival of the current infant.

Parity - The number of infants produced by a female.

Paternal investment theory - Parents will be selected to invest their resources in a way that will function to maximize their lifetime inclusive fitness.

Pentadactyly - Possession of five digits on the hands and feet.

Perineal skin - The skin covering the area surrounding the openings of the anus, vagina and urethra.

Permissive - A female who allows conspecifics to caretake her infants.

Phenotype - The observable appearance of an organism which is determined by the interaction of the genotype and environmental factors during development.

Pheremone - A chemical released from secretory glands which functions as a communicator between conspecifics (see releasor and primer pheremones).

Philopatry - Strictly speaking this refers to love of country, but in primate studies it describes a situation in which an individual or group of individuals remain in their natal group for life.

Phylogeny - The genealogical relationship between groups of organisms.

Placental mammals - Organisms in which the infant develops inside th body of the pregnant female. The placenta is the organ that supplies oxygen and nourishment to the developing fetus.

Platyrrhini - New World monkeys. The term refers to their flat, side-directed nostrils.

Poligamy - Mating pattern in which the members of one sex mate with more than one member of the other sex.

Polyandry - Mating pattern in which one female mates with more than one male.

Polygamy - Mating pattern in which an indivual has more than one mate.

Polygyny - Mating pattern in which one male mate mates with more than one female.

Polyspecific groups - Groups of animals made up of more than one species.

Postcranial - Referring to the skeletal parts below the neck.

Post-orbital bar - Bony ring surrounding the eye sockets.

Power grip - The forceful grip shown by some primate species when grasping an object between the underside of the mobile digits and the palm.

Precision grip - The ability to pick up small objects between the thumb and forefinger provided by the manipulative digits and the opposable thumbs found in many primate species, particularly Old World monkeys.

Precocial - Independant at an early age.

Prehensile - Grasping.

Primer pheromone - A chemical communicator which affects the recipient over a period time.

Primiparous- A female who has only borne one infant.

Primitive - An ancestral behavioral or physical charcteristic that is found in the modern species.

Proceptivity - A state in which a female is motivated to mate which is often characterized by active solicitation of sexual behavior.

Prognathic - Protruding.

Proximate - Immediate.

Quadrumanual - Locomotor pattern in which both the hands anfeet are used to move through the trees.

Quadrupedal - Locomotor pattern characterized by movement along the substrate using all four limbs equally.

Reciprocal Altruism Theory - An individual will aid another, while putting itself at risk, if a return of the favor is expected at a later date.

Red Data Book - Major reference for endangered plants and animals, published by the I.U.C.N.

Releasor pheromones - Chimical communicators that produce an immediate reaction in the recipient.

Retinal Cones - Photoreceptor cells that function in high intensity light and provide the basis for color vision.

Retinal fovea - The light sensitive area of the retina

Saggital Crest - A bony ridge running along the midline of the skull that functions as an attachment area for the jaw muscles.

Saltatory - Leaping style of locomotion.

Scent marking - Olfactory messages that are dispersed into the air or deposited on the substrate.

Semi-terrestrial - Spending part of the time on the ground and part of the time in the trees.

Series mounters - Species that require the male to mount the female several times in succession over a period of about 20 minutes for a successful copulation to occur.

Sexual dimorphism - The differences between the males and females of a given species.

Silverbacks - Adult male gorillas, who exhibit a saddle of gray hairs across their backs when they reach 13 to 14 years of age.

Social Group - Animals who interact on a regular basis, able to recognize each other and spend more time with group members than other conspecifics.

Social Organization - Refers to the spatial distribution, composition, and social and physical relationships within the group. Differs from social structure in that it includes a behavioral component.

Social Structure - The physical make-up of a group in terms of numbers of animals, sex ratios and relationships between group members.

Sociobiology - The study of the biological basis for social behavior.

Socioecology - The study of the relationship between the environment and social patterns.

Sonogram - A spectrographic picture of sound waves.

Stereoscopic Vision - The ability to see objects in three dimensions produced the superimposition of two images on the visual cells in the brain.

Substrate - The area of the environment (ground, branch etc.) on which an animal moves, attaches or rests.

Supplantation - An individual moves away from its spatial position and avoids a confrontation when approached by another individual.

Sympatric - Species that range in the same geographical area.

Tactile - The sense of touch.

Tapetum lucidum - A crystalline sheild behind the retina that functions to reflect light onto the retina. Commonly found in nocturnal animals.

Taxa - A unit of classification.

Taxonomy - The science of the classification of organisms.

Terrestrial - Living primarily on the ground.

Territory - The area of a groups' home range that is actively defended against intruders.

Testosterone - A male hormone.

Transitivity - The concept that: if A is greater than B, and B is greater than C, then it follows that A is greater than C.

Tumescence - Swelling.

Ultimate - Evolutionary.

Vertical Clinging and Leaping - A locomotor pattern characterized by leaping through the trees and clinging to the branches with the upper body erect.

Xenophobic alliance - A coalition of conspecifics acting against strangers, either of the same or of a different species.

Y 5 cusp pattern - A dental pattern in which the lower molars have five cusps arranged so that the fissures between the cusps form the letter Y. This dental characteristic is unique to hominoids.

Alexander, R.D. 1974. The evolution of social behavior. *Annual Review of Ecology and Systematics*, 5, 325–383.

Anderson, N.D. and W.R. Brown. 1984. *Lemurs*. Dodd, Mead, New York, N.Y.

Antinucci, F. 1989. *Cognitive Structure and Development in Nonhuman Primates*. Hillsdale, New Jersey.

Baldwin, J.D. 1985. Behavior of squirrel monkeys (*Saimiri*) in natural environments. In *Handbook of Squirrel Monkey Research*, L.A. Rosenblum and C.L. Coe eds. Plenum Press, New York, N.Y. pp. 33–50.

Beck, B.B. 1980. *Animal Tool Behavior*. Garland Press, New York, N.Y.

Berman, C.M. 1984. Variations in mother-infant relationships: Traditional and non-traditional factors. In *Female Primates: Studies by Female Primatologists*, M.E. Small ed. Alan Liss Inc. New York, N.Y. pp. 17–36.

Bernstein, I.S. 1964. The integration of rhesus monkeys introduced to a group. *Folia Primatologica* 2:50–63.

Bernstein, I.S. 1967. Field study of the pigtail macaques (*Macaca nemestrina*). *Primates* 8:217–228.

Burton, F. 1984. Inference of cognitive abilities in Old World monkeys. *Semantica* 50:69–81.

Chapais, B. 1991. Primates and the origins of aggression, power and politics among humans. In *Understanding Behavior*, J.D. Loy and C.B. Peters eds. Oxford University Press, New York, N.Y. pp. 190–228.

Cheney D.L. and R.M. Seyfarth. 1980. Vocal recognition in free-ranging vervet monkeys. *Animal Behaviour* 38:362–367.

Cheney, D.L. and R.M. Seyfarth. 1990. *How Monkeys See the World*. University of Chicago Press.

Cheney, D.L., R.M. Seyfarth, B.B. Smuts, and R.W. Wrangham. 1987. The study of primate societies. In *Primate Societies*, B.B. Smuts, D.L. Cheney, R.M. Seyfarth, T.T. Struhsaker and R.W. Wrangham eds. University of Chicago Press, Chicago, Ill. pp. 1–8.

Chism, J. T. Rowell and D. Olson. 1984. Life history patterns of female patas monkeys. In *Female Primates: Studies by Female Primatologists*, M.E. Small ed. Alan Liss Inc. New York, N.Y. pp. 175–189.

Ciochon, R.I. and A.B. Chiarelli. 1980. *Evolutionary Biology of the New World Monkeys and Continental Drift*. Plenum Press, New York, N.Y.

Clark, W.E. Le Gros. 1959. *The Antecedents of Man*. University Press, Edinburgh, U.K.

Clutton-Brock, T.H. ed. 1977. *Primate Ecology: Studies of Feeding and Ranging Behaviour in Lemurs, Monkeys and Apes*. Academic Press, London, U.K.

Collinge, N.E. 1987. Weaning variability in semi-free ranging Japanese macaques (*Macaca fuscata*). *Folia Primatologica* 48:137–150.

Collinge, N.E. 1991. Variability in aspects of the mother-infant relationship in Japanese macaques during weaning. In *The Monkeys of Arashiyama*, L.M. Fedigan and P.L Asquith eds. pp. 157–174.

Cords, M. 1987. Forest guenons and patas monkeys: Male-male competition in one-male groups. In *Primate Societies*, B.B. Smuts, D.L. Cheney, R.M. Seyfarth, T.T. Struhsaker and R.W. Wrangham eds. University of Chicago Press, Chicago, Ill. pp.98–111.

Costa, J.P.O. 1987. Brazil: Conservation Policy. In *For the Conservation of Earth*, V. Martin ed. Fulcrum Inc., Golden, Colorado. pp. 176–180.

Crockett, C.M. and J.F. Eisenberg. 1987. Howlers: Variations in group size and demography. In *Primate Societies*, B.B. Smuts, D.L. Cheney, R.M. Seyfarth, T.T. Struhsaker and R.W. Wrangham eds. University of Chicago Press, Chicago, Ill, pp. 54–68.

Crompton, R.H. and P.M. Andau. 1987. Ranging activity and sociality in free-ranging *Tarsius bacanus*: A preliminary report. *International Journal of Primatology* 8:43–72.

Crook, J.H. and J.S. Gartlan. 1966. The evolution of primate societies. *Nature* 210:1200–1203.

Darwin, C. 1859. *The Origin of the Species, and the Descent of Man*. Random House, New York, N.Y.

Darwin, C. 1871. *The Descent of Man and Selection in Relation to Sex*. D. Appleton, New York, N.Y.

Demment, M.W. 1983. Feeding ecology and the evolution of brain size in baboons. *African Journal of Ecology* 21:219–233.

de Waal, F.M.B. 1987. Dynamics of social relationships. In *Primate Societies*, B.B. Smuts, D.L. Cheney, R.M. Seyfarth, T.T. Struhsaker and R.W. Wrangham eds. University of Chicago Press, Chicago, Ill, pp. 421–430.

de Waal, F.M.B. 1989. *Peacemaking Among Primates*. Harvard University Press, Cambridge, Mass.

Dickinson, R.E. 1981. Effects of tropical deforestation on climate. In *Blowing in the Wind: Deforestation and Long-range Implications*, V.H. Sutlive, N.A. Altshuler, M.D. Zamora eds. Department of Anthropology, College of William and Mary, Williamsberg, VA. pp.411–442.

Drickamer, L.C. and S.H. Vessey. 1982. *Animal Behavior: Concepts, Processes and Methods*. P.W.S. Publishers, Boston, Mass.

Dunbar, R.I.M. and E.P. Dunbar. 1975. *Social Dynamics of Gelada Baboons*, Karger, Basil, Switzerland.

Eisenberg, J.F. N.A. Muckenhirn, and R. Rudran. 1972. The relation between ecology and social structure in primates. *Science* 176: 863–874.

Essock-Vitale, S. and R.M. Seyfarth. 1987. Intelligence and social cognition. In *Primate Societies*, B.B. Smuts, D.L. Cheney, R.M. Seyfarth, T.T. Struhsaker and R.W. Wrangham eds. University of Chicago Press, Chicago, Ill, pp. 452–461.

Eudy, A.A. 1987. *Action Plan for Asian Primate Conservation 1987–91. IUCN/SSC Primate Specialist Group*.

Fedigan, L and L.M. Fedigan. 1988. *Cercopithecus aethiops*: A review of field studies. In *A Primate Radiation: Evolutionary Biology of the African Guenons*, A Gauthier-Hion, F. Bourliere, J.P. Gauthier and J. Kingdon eds. Cambridge University Press, Cambridge, U.K. pp. 389–411.

Fedigan, L.M. 1990. Vertebrates predation in *Cebus capucinus*: Meat eating in a neo-tropical monkey. *Folia Primatologica* 54:196–205.

Fedigan, L.M. 1992. *Primate Paradigms:Sex Roles and Social Bonds* (2nd Ed). University of Chicago Press, Chicago. Ill.

Flavin, C. 1988. The heat is on. *World Watch* 1:10–20

Fleagle, J.G. *Primate Adaptation and Evolution*. Academic Press Inc. New York, N.Y.

Fontaine, R. 1981. The uakaris, genus *Cacajao*. *In The Ecology and Behavior of Neotropical Primates. Vol. 1.*, A.F. Coimbra-Filho and R.A. Mittermeier eds. Academia Brasileira de Ciencias, Reo de Janeiro,

Fouts, R.S. 1973. Acquisition and testing of gestural signs in four young chimpanzees. *Science* 180:778–780.

Fouts, R.S. 1983. Chimpanzee language and elephant tails: A theoretical synthesis. In *Language in Primates: Perspectives and Implications.* J deLuce and H.T. Wilder eds. Springer-Verlag, New York. pp.-63–70.

Fragaszy, D.M. 1985. Cognition in squirrel monkeys (*Saimiri*). A contemporary perspective. In *Handbook of Squirrel Monkey Research*, L.A. Rosenblum and C.L. Coe eds. Plenum Press, New York, N.Y. pp.55–98.

Fragaszy D.M. and G. Mitchell. 1974. Infant socialization in primates. *Journal of Human Evolution* 3:563–574.

Galdikas, B.M.F. 1979. Orangutan adaptation at Tanjung Puting Reserve: Mating and ecology. In *The Great Apes*, D.A Hamburg and E.R. McCown eds. Benjamin/ Cummings, Menlo Park. pp. 195–233.

Galdikas, B.M.F. 1981. Orangutan reproduction in the wild. In *The Reproductive Biology of the Great Apes*, C.E. Graham ed. Academic Press, New York, N.Y. pp. 281–300.

Galdikas, B.M.F. 1982. Orangutan tool use at Tanjung Puting Reserve, Central Indonesian Borneo (Kalimantang Tengah). *Journal of Human Evolution* 11:19–33.

Gardner, B.T. and R.A. Gardner. 1979. Two comparative psychologists look at language acquisition. In *Childrens' Language, Vol.2*. K.E. Nelson ed. Halstead, New York, N.Y. pp. 309–369.

Gardner, R.A. and B.T. Gardner. 1969. Teaching sign language to a chimpanzee. *Science* 165:664–672.

Gibson, K.R. 1986. Cognition, brain size and the extraction of embedded food. In *Primate Ontogeny:Cognitive and Social Behavior*, J.G. Else and P.C. Lee eds. Cambridge University Press, Cambridge, Mass. pp. 93–104.

Gillan, D.J. 1981. Reasoning in a chimpanzee: II. Transitive inference. *Journal of Experimental Psychology* 7:150–164.

Gillan, D.J., D. Premack, and G Woodruff. 1981. Reasoning in a chimpanzee: I Analogical reasoning. *Journal of Experimental Psychology* 7:1–17.

Goodall, J. 1965. Chimpanzees of the Gombe Stream Reserve. In *Primate Behavior: Field Studies of Monkeys and Apes*, I. deVore ed. Holt, Rinehart and Winston, New York, N.Y. pp. 425–473.

Goodall, J. 1968. A preliminary report on expressive movements and communication in the Gombe Stream chimpanzees. In *Primates: Studies in Adaptation and Variability*, P.E. Jay ed. Holt, Rinehart and Winston, New York, N.Y. pp. 313–374.

Goodall, J. 1986. *The Chimpanzees of Gombe: Patterns of Behavior*. Harvard University Press, Cambridge, Mass.

Goodall, J. 1990. *Through a Window: My Thirty Years With the Chimpanzees*. Houghton Mifflin Co., Boston, Mass.

Goy, R.W. 1966. Role of androgen in the establishment and regulation of behavioral sex differences in mammals. *Journal of Animal Science* 25: Supplement 21–35.

Goy, R,W. 1968. Organizing effects of androgen on the behaviors of rhesus macaques. In *Endocrinology and Human Behavior*, R.P. Michael ed. Oxford University Press, London. pp. 12–31.

Gouzoules H. 1980. A description of genealogical rank changes in a troop of Japanese monkeys (*Macaca fuscata*). *Primates* 21:262–267.

Gouzoules, S. 1984. Primate mating systems, kin associations and cooperative behavior: Evidence of kin recognition? *Yearbook of Physical Anthropology* 27: 99–134.

Gouzoules, S. and H. Gouzoules. 1987. Kinship. In *Primate Societies*, B.B. Smuts, D.L. Cheney, R.M. Seyfarth, T.T. Struhsaker and R.W. Wrangham eds. University of Chicago Press, Chicago, Ill, pp. 299–305.

Grainger, A. 1987. Tropical rain forests: Global resources of national responsibility. In *For the Conservation of Earth*, V. Martin ed. Fulcrum Inc. Golden, Colorado. pp. 94–104.

Gray, J.P. 1985. *Primate Sociobiology*. HRHF Press, New Haven, Conn.

Green, S. 1981. Sex differences in age gradations in vocalizations of Japanese and lion-tailed macaques (*Macaca fuscata* and *Macaca silenus*) *American Zoologist* 21: 165–183.

Greenfield, P.M. and E.S. Savage-Rumbaugh. 1990. Grammatical combinations in *Pan paniscus*: Processes of learning and invention in the evolution and development of language. In *"Language" and Intelligence in Monkeys and Apes*, S.T. Parker and K.R. Gibson eds. Cambridge University Press, Cambridge, Mass. pp. 540–578.

Griffin, D.R. *The Question of Animal Awareness*. Rockerfeller University Press, New York. N.Y.

Hall, K.R.L. 1968. Social living in monkeys. In *Primates: Studies in Adaptation and Variability*, P.C.Jay ed. Holt, Rinehart and Winston, New York. pp.383–397.

Hamilton, W.D. 1963. The evolution of altruistic behavior. *American Naturalist* 97:354–356.

Haraway, D. 1989. *Primate Visions*. Routledge, Chapman and Hall Inc., New York, N.Y.

Harcourt, A.H. 1988. Alliances and social intelligence. In *Machiavellian Intelligence. Social Expertise and the Evolution of Intellect in Monkeys, Apes and Humans*, R. Byrne and A. Whiten eds. Clarendon Press, Oxford, U.K. pp. 132–152.

Harcourt, A.H. 1992. Coalitions and alliances: Are primates more complex than non-primates. In *Coalitions and Alliances in Humans and Other Animals*, A.H. Harcourt and F.B.M. deWaal eds. Oxford University Press, New York, N.Y. pp. 445–472.

Harlow, H.F. 1949. The formation of learning sets. *Psychological Review* 56:51–65.

Harlow, H.F. 1965. Sexual behavior in rhesus monkeys. *American Psychologist* 17:1–9.

Harlow, H.F, and M.K. Harlow. 1962. Social deprivation in monkeys. *Scientific American* 207:136–146.

Harlow H.F., M.K. Harlow and E.W. Hansen. 1963. The maternal affectional system of rhesus monkeys. In *Maternal Behavior in Mammals*, H. Reingold ed. John Wiley and Sons, New York, N.Y. pp. 254–281.

Hayes, C. 1951. *The Ape in Our House*. Harper, New York, N.Y.

Hinde, R.A. 1974. Introduction. In *The Biological Bases of Human Social Behaviour*. McGraw-Hill Book Co., New York, N.Y. pp.1–6.

Hinde, R.A. and Y. Spencer-Booth. 1970. Individual differences in the responses of rhesus monkeys to a period of separation from their mothers. *Journal of Child Psychiatry* 11:159–176.

Hrdy, S.B. 1977a. Infanticide as a primate reproduction strategy. *American Scientist* 65:40–49.

Hrdy, S.B. 1977b. Male-male competition and infanticide among the langurs (*Presbytis entellus*) of Abu, Rajasthan. *Folia Primatologica* 22:19–58.

Hrdy, S.B. and P.L. Whitten. 1987. Patterning of sexual activity. In *Primate Societies*, B.B. Smuts, D.L. Cheney, R.M. Seyfarth, T.T. Struhsaker and R.W. Wrangham eds. University of Chicago Press, Chicago, Ill, pp. 370–384.

Humphrey, N.K. 1976. The social function of intellect. *In Growing Points in Ethology*, P.P.G. Bateman and R.A. Hinde eds. Cambridge University Press, Cambridge, U.K. pp. 303–319.

Humphreys, J. 1982. Plants that bring health or death. *New Scientist* Feb. 25 pp.513–516.

Itani, J. and A. Nishimura. 1973. The study of infrahuman culture in Japan. In *Precultural Primate Behavior, Symposium of the 4th Congress of the International Society of Primatologists*, Volume 1, E.W. Menzel ed. S. Karger, Basil, Switzerland. pp. 26–50.

Iwano, T. and C Iwakawa. 1988. Feeding behavior of the aye-aye (*Daubentonia madagascariensis*). *Folia Primatologica* 50:136–142.

Izawa, K. 1979. Foods and feeding behavior of wild black-capped capuchins (*Cebus apella*). *Primates* 20:57–76.

Jerison, H.J. *The Evolution of Brain and Intelligence*. Academic Press, New York, N.Y.

Jolly, A. 1966. Lemur social behavior and primate intelligence. *Science* 153: 501–506.

Jolly, A. 1985. The *Evolution of Primate Behavior. Second Edition*. Macmillan Publishing Co. New York, N.Y.

Jolly, A. 1988. The evolution of purpose. In *Machiavellian Intelligence:Social Expertise and the Evolution of Intelllect in Monkeys, Apes and Humans*, R. Byrne and A. Whiten eds. Clarendon Press, Oxford, U.K. pp. 363–378.

Jouventin, P. 1975. Observations sur la socio-ecologie au mandrill. *Terre et Vie* 29:493–532.

Kano, T. 1980. The social behavior of wild pygmy chimpanzees (*Pan paniscus*) of Wamba. A preliminary report. *Journal of Human Evolution* 9:243–260.

Kavanagh, M. 1972. Food sharing behavior within a group of douc monkeys (*Pygathrix nemaeus*). *Nature* 239:406–407.

Kavanagh, M. 1983. *A Complete Guide to Monkeys, Apes and Other Primates*. Cape Press, London, U.K.

Kawai, M. 1965. Newly acquired precultural behavior of the natural troop of Japanese monkeys on Koshima Island. *Primates* 6:1–30.

Kawamura, S. 1959. The process of sub-culture propagation among Japanese macaques. *Primates* 2:43–54.

Kellogg, W.N. and L.A. Kellogg. 1933. *The Ape and the Child:A Study of Environment Influence and Behavior*. Heffner, New York, N.Y.

Kingdon, J. 1974. *East African Mammals:Volume 1*. University of Chicago Press, Chicago, Ill.

Kleiman, D. 1980. The sociobiology of captive propagation. In *Conservation Biology: An Evolutionary-Ecological Perspective*. M.E. Soule and B.A. Wilcox eds. Sinauer Associates, Sunderland, Mass. pp. 243–261

Kluver, H. 1937. *Behavioral Mechanisms in Monkeys*. University of Chicago Press, Chicago, Ill.

Kummer, H. 1968. *Social Organization of Hamadryas Baboons*. University of Chicago Press, Chicago, Ill.

Kummer, H. 1982. Social Knowledge in free-ranging primates. In *Animal Mind-Human Mind*, D.R. Griffin ed. Springer-Verlag, New York, N.Y. pp 113–130.

Leighton, D.R. 1985. Gibbons: Territoriality and monogamy. In *Primate Societies*, B.B. Smuts, D.L. Cheney, R.M. Seyfarth, T.T. Struhsaker and R.W. Wrangham eds. University of Chicago Press, Chicago, Ill, pp.135–145.

leBoef, B.J. 1974. Male-male competition and reproductive success in elephant seals. *American Zoologist* 14:163–176.

Lehner, P.N. 1979. *Handbook of Ethological Methods*. Garland STPM Press, N.Y.

Lindon, E. 1986. *Silent Partners*: The Legacy of Ape Language Experiments. Time Books, New York, N.Y.

Lindon, E. 1992. A curious kinship: Apes and humans. *National Geographic* Mar. pp.3–45.

Lorenz, K. The comparative method of studying animal behaviour patterns. *Symposium of Social Experiences in Biology* 4:221–268.

Lorenz, K. 1966. *On Aggression*. Metheuen, London, U.K.

Lorenz, K. 1970. *Studies in Animal and Human Behaviour. Vol.1*. Metheuen, London.

Mack, D. and R.A. Mittermeier. 1984. The international primate trade: Summary, update and conclusions. In *The International Primate Trade. Vol I.*, D. Mack and R.A. Mittermeier eds. Traffic (USA) Washington, D.C.

Mallinson, J.J.C. 1986. International efforts to secure a viable population of the golden-headed lion tamarin. *IUCN Newsletter* 8:124–126.

Marsh, C.W. 1981. Time budget of Tana River red colobus. *Folia Primatologica* 35:30–50.

Martin R.D., G.A. Doyle and A.C. Walker. 1974. *Prosimian Biology*. Gerald Duckworth and Co. Ltd, London, U.K.

Mason, W.A. 1975, Comparative studies of social behavior in *Callicebus* and *Saimiri*:Behavior of male and female pairs. *Folia Primatologica* 22:1–8.

Mason, W.A. Early social deprivation in the nonhuman primate: Implications for human behavior. In *Environmental Influences*, D.L. Glass ed. Rockerfeller University Press, New York, N.Y. pp.70–100.

McGonigle, B.O. and M. Chalmers 1977. Are monkeys logical? *Nature* 267:694–696.

McKenna, J.J. 1979. Aspects of infant socialization, attachment and maternal caregiving patterns among primates: A cross-disciplinary review. *Yearbook of Physical Anthropology* 22:250–286.

Melnick D.J. and M.C. Pearl 1987. Cercopithecines in multimale groups: Genetic diversity and population structure. In *Primate Societies*, B.B. Smuts, D.L. Cheney, R.M. Seyfarth, T.T. Struhsaker and R.W. Wrangham eds. University of Chicago Press, Chicago, Ill, pp. 121–134.

Mendoza, S.P. 1984. The psychology of social relationships. In *Social Cohesion: Essays*

Towards a Sociopsychological Perspective, P.R Barches and S.P. Mendoza eds. Greenwood Press, Westport, Conn. pp. 3–23.

Menzel, E.W. 1972. Spontaneous investigation of ladders in a group of young chimpanzees. *Folia Primatologica* 15:220–232.

Menzel, E.W. 1974. A group of young chimpanzees in a one-acre field. In *Behavior of Non-Human Primates, Vol. 5*, A.M. Schrier and F. Stollnitz eds. Acacdemic Press, New York, N.Y. pp. 83–153.

Michael R.P. and D. Zumpe, 1971. Patterns of reproductive behavior. In *Comparative Reproduction in Nonhuman Primates*, E.S.E. Hafez ed. C.C. Thomas, Springfield, Mass. pp. 205–242.

Miles, H.L. 1983. Apes in language: The reach for communicative competence in *Language in Primates: Perspectives and Implications*, J. deLuce and H.T, Wilder eds. Springer-Verlag, New York, N.Y. pp. 43–62.

Miles H.L.W. 1990. The cognitive foundations for reference in a signing orangutan. In *"Language" and Intelligence in Monkeys and Apes*, S.T. Parker and K.R. Gibson eds. Cambridge University Press, Cambridge, Mass. P.. 511–539.

Milton, K. 1988. Foraging behavior and the evolution of primate intelligence. In *Machiavellian Intelligence:Social Expertise and the Evolution of Intelllect in Monkeys, Apes and Humans*, R. Byrne and A. Whiten eds. Clarendon Press, Oxford, U.K. pp. 285–306.

Mitchell, G.D., W.K. Redican,and J. Gombe. 1974. Lessons from a primate: males can raise babies. *Psychology Today* 7:63–70.

Mittermeier, R.A. and D.L. Cheney. 1987. Conservation of primates and their habitats. In *Primate Societies*, B.B. Smuts, D.L. Cheney, R.M. Seyfarth, T.T. Struhsaker and R.W. Wrangham eds. University of Chicago Press, Chicago, Ill, pp. 477–490.

Mittermeier, R.A., C.M.C. Valle, M.C. Alves, I.B. Santos, C.A.M. Pinto, K. Strier, A.L. Young, E.M. Veado, I.D. Constable, S.G. Paganella and R.M. Lemos de Sa. 1987. Current distribution of the muriqui in the Atlantic forest region of Eastern Brazil. *IUCN Newsletter* 8:143–144.

Morgan, C.L. 1896. *Introduction to Comparative Psychology*. Scott, London, U.K.

Morris, R and D. Morris. 1968. *Men and Apes*. Sphere Books Ltd., London, U.K.

Murray, R.D. and K.M. Murdoch. 1987. Mother-infant dyad behavior in the Oregon troop of Japanese monkeys. *Primates* 18:15–24.

Myers, N. 1984. *The Primary Source: Tropical Forests and Our Future*. W.W. Norton, New York, N.Y.

Napier, J.R. and P.H. Napier. 1967. *Handbook of Living Primates*. Academic Press, London, U.K.

Napier, J.P. and P.H. Napier. 1985. *The Natural History of Primates*. M.I.T. Press, Cambridge, Mass.

Nicolson, N.A. 1987. Infants, mothers and other females. In *Primate Societies*, B.B. Smuts, D.L. Cheney, R.M. Seyfarth, T.T. Struhsaker and R.W. Wrangham eds. University of Chicago Press, Chicago, Ill, pp. 330–342.

Nishida, T. 1970. Social behavior and relationships among wild chimpanzees of the Mahale mountains. *Primates* 11:47–87.

Nishida, T. 1987. Local tradition and cultural transmission. In *Primate Societies*, B.B. Smuts,

D.L. Cheney, R.M. Seyfarth, T.T. Struhsaker and R.W. Wrangham eds. University of Chicago Press, Chicago, Ill, pp. 462–470.

Nishida, T. and M. Hiraiwa-Hasegawa. 1987. Chimpanzees and bonobos cooperative relationships among males. In *Primate Societies*, B.B. Smuts, D.L. Cheney, R.M. Seyfarth, T.T. Struhsaker and R.W. Wrangham eds. University of Chicago Press, Chicago, Ill, pp. 165–178.

Oates, J.F. 1985. *Action Plan for African Primate Conservation: 1986–90*. IUCN/SSC Primate Specialist Group.

Oko, R.A. 1992. The present situation of conservation for wild gorillas in the Congo. In *Topics in Primatology Vol.2.*, N. Itoigawa, Y. Sugiyama, G.P. Sackett and R.K.R. Thompson eds. University of Tokyo Press, Tokyo, Japan. pp. 241–244.

Oren, D.C. 1987. Grande Carajas, International financing agencies and biological diversity in Southeastern Brazilian Amazonia. *Conservation Biology* 1:222–227.

O'Rourke, D.H. 1979. Components of genetic and environmental variation in hamadryas baboon morphometrics. *American Journal of Physical Anthropology* 50:469.

Packer, C. 1977. Reciprocal altruism in Papio anubis. *Nature* 265:441–443.

Packer, C. 1979. Intertroop transfer and inbreeding in *Papio anubis*. *Animal Behaviour* 27:37–45.

Parker, S.T. and K.R. Gibson. 1977. Object manipulation, tool use and sensorimotor intelligence in feeding adaptations in cebus monkeys and great apes. *Journal of Human Evolution* 6:623–641.

Patterson, F.C. and E. Linden.1981. *The Education of Koko*. Holt, Rinehart and Winston, New York, N.Y.

Pereira, M.E., M.L. Seligman, and J.M. Macedonia. 1988. The behavioral repertoire of black and white ruffed lemurs, *Varecia variegata* (Primate:Lermuridae). *Folia Primatologica* 51:1–32.

Pereira, M.E., R. Kaufman, P.M. Keppeler, D.J. Overdorff. 1990. Female dominance doesn't characterize all of the Lorisidae. *Folia Primatologica* 55:96–103.

Peterson, D. 1989. *The Deluge and the Ark: A Journey Into Primate Worlds*. Avon Books, New York.

Poirier, F.E. 1972. Introduction. In *Primate Socialization*, F.E. Poirier ed. Random House, New York N,Y. pp. 3–28.

Poti, P and F. Antinucci. 1989. Logical operations. In *Cognitive Structure and Development in Nonhuman Primates*, F. Antinucci ed. Laurence Erlbaum Associates, Hillsdale, New Jersey, pp. 189–227.

Premack D. and G. Woodruff. 1978. Does a chimpanzee have a theory of mind? *Behavioral and Brain Sciences* 1:515–526.

Premack D. and A.J. Premack. 1983. *The Mind of an Ape*. W.W. Norton and Co. New York, N.Y.

Ralls, K. and J. Ballou. 1982. Effects of inbreeding on infant mortality in captive primates. *International Journal of Primatology* 3:491–505.

Redican, W.K. 1976. Adult male-infant interactions in non-human primates. In *The Role of the Father in Child Development*, M.E. Lamb ed. John Wiley and Sons, New York, N.Y.pp. 345–485

Richard, A. 1985. *Primates in Nature*. W.H. Freeman and Co.

Ristau, C.A. and D. Robbins. 1982. Language in great apes: A critical review. In *Advances in the Study of Behavior*, J.S. Rosenblatt, R.A. Hinde, C.Beer and M.C. Busnel eds. Plenum Press, New York, N.Y. pp. 140–255.

Rodman, P.S. and J.C. Mitani. 1987. Orangutans: Sexual dimorphism in a solitary species. In *Primate Societies*, B.B. Smuts, D.L. Cheney, R.M. Seyfarth, T.T. Struhsaker and R.W. Wrangham eds. University of Chicago Press, Chicago, Ill., pp.146–154.

Romanes, G. 1882. *Animal Intelligence*. Kegan Paul, London, U.K.

Rose, R.M., T.P. Goroq, and I.S. Bernstein. 1972. Plasma testosterone levels in the male rhesus: Influences of sexual and social stimuli. *Science* 178:643–645.

Rumbaugh, D.M. and T.V. Gill. 1977. Lana's acquisition of language skills. In *Language Learning by a Chimpanzee: The Lana Project*, D.M. Rumbaugh ed. Academic Press, New York, N.Y. pp. 165–192.

Sackett, G.P. 1974. Sex differences in rhesus monkeys following varied rearing experiences. In *Sex Differences in Behavior*, R.C. Friedman, R.M. Richart and R.L. Vande Wiele eds. John Wiley and Sons, New York, N.Y. pp.99–122.

Sade, D.S. 1968. Inhibition of mother-son mating among free-ranging rhesus monkeys. *Science and Psychoanalysis* XII:18–38.

Sade, D.S. 1991. Kinship. In *Understanding Behavior*, J.D. Loy and C.B. Peters eds. Oxford University Press, New York, N.Y. pp. 229–241.

Santos, I.B., R.A. Mittermeier, A.B. Rylands, and C Valle. 1987. The distribution and conservation status of primates in Southern Bahia, Brazil. *IUCN Newsletter* 8:126–143.

Savage-Rumbaugh, E.S., D.M. Rumbaugh and S. Boysen. 1978. The symbolic communication between two chimpanzees (*Pan troglodytes*). *Science* 201:641–644.

Sawaguchi, T and H. Kudo. 1990. Neocortical development and social structure. *Primates* 31:283–289.

Schaller, G.B. 1965. The behavior of the mountain gorilla. In *Primate Behavior: Field Studies of Monkeys and Apes*, I. deVore ed. Holt, Rinehart and Winston, New York, N.Y. pp. 324–367.

Schjelderup-Ebbe, T. 1922. Beitrage zur sozialpsychologie das haushuhns. *Zeitschrift fur Psychologie* 88:225–252.

Scott, J.P. 1966. Agonistic behavior of mice and rats: A review. *American Zoologist* 6:683–701.

Scott, J.P. 1974. Agonistic behavior of primates: A comparative perspective. In *Primate Aggression, Territoriality and Xenophobia*, R. Holloway ed. Academic Press, N.Y. pp 417–434.

Seboek, T.A. and J. Umiker-Seboek. 1980. Questioning apes. In *Speaking of Apes*, T.A. Seboek and J. Umiker-Seboek eds. Plenum Press, New York, N.Y. pp 429–440.

Seyfarth, R.M., D.L. Cheney and P. Marler. 1980. Monkey responses to three different alarm calls. Evidence for predator classification in semantic communication. *Science* 210:801–803.

Sitwell, N. 1988. Monkey see, monkey pick. *International Wildlife*. May-June pp. 18–21.

Skinner, B.F. 1938. *The Behavior of Organisms: An Experimental Analysis*. Appleton-Century-Crofts, New York, N.Y.

Smith, W.J. 1977. *The Behavior of Communicating*. Harvard University Press, Cambridge, Mass.

Smuts, B.B. 1985. *Sex and Friendship in Baboons*. Aldine Press, New York.

Smuts, B.B. 1987a. Sexual competition in mate choice. In *Primate Societies*, B.B. Smuts, D.L. Cheney, R.M. Seyfarth, T.T. Struhsaker and R.W. Wrangham eds. University of Chicago Press, Chicago, Ill. pp. 385–399.

Smuts, B.B. 1987b. Gender, aggression and influence. In *Primate Societies*, B.B. Smuts, D.L. Cheney, R.M. Seyfarth, T.T. Struhsaker and R.W. Wrangham eds. University of Chicago Press, Chicago, Ill. pp.400–412.

Southwick C.H. and R.B. Smith. 1986. The growth of primate field studies. In *Comparative Primate Biology. Vol. 2A: Behavior, Conservation and Ecology*, G. Mitchell and J. Erwin eds. Alan R.Liss, New York, N.Y. pp. 73–91.

Stammbach, E. 1987. Desert, forest and montane baboons: Multilevel societies. In *Primate Societies*, B.B. Smuts, D.L. Cheney, R.M. Seyfarth, T.T. Struhsaker and R.W. Wrangham eds. University of Chicago Press, Chicago, Ill., pp. 112–120.

Stephan, H. 1972. Evolution of the primate brain: Comparative anatomical investigation. In *Functional and Evolutionary Biology of Primates*, R. H.Tuttle ed. Maine Press, Chicago, Ill. pp. 155–174.

Stewart K.J. and A.H. Harcourt. 1987. Gorilla female friendships. In *Primate Societies*, B.B. Smuts, D.L. Cheney, R.M. Seyfarth, T.T. Struhsaker and R.W. Wrangham eds. University of Chicago Press, Chicago, Ill., pp. 155–164.

Struhsaker, T.T. 1981. Polyspecific associations among tropical forest primates. *Zeitschrift fur Tierpsychologie*. 57:268–304.

Struhsaker, T.T. and J.S. Gartlan. 1970. Observations on the behavior and ecology of the patas monkey (*Erythrocebus patas*) in the Wasa Reserve, Cameroon. *Journal of Zoology (London)* 161:49–63.

Struhsaker, T.T. and J.F. Oates. 1975. Comparison of the behavior and ecology of the red colobus and black and white colobus monkeys in Uganda: A summary. In *Socioecology and Psychology of Primates*, R.H. Tuttle ed. Aldine Publishing Co., Chicago, Ill. pp. 103–123.

Strum, S.C. *Almost Human: A Journey into the World of Baboons*. Random House, New York, N.Y.

Sugiyama, Y. 1965. On a social change of Hanuman langurs (*Presbytis entellus*) in their natural condition. *Primates* 6:381–418.

Tarpy, R.M. 1982. *Principles in Animal Learning and Motivation*. Scott, Forseman and Co., Glenview, Ill.

Tattersall, I. 1993. Madagascar's lemurs. *Scientific American*. January 1993 pp. 110–117.

Terborgh, J. *Five New World Primates: A Comparative Perspective*. Princeton University Press, New Jersey.

Terborgh, J. and A.W. Goldizen. 1985. On the mating system of the cooperatively feeding saddle-back tamarin (*Saguinis fuscicollis*). *Behavior, Ecology and Sociobiology* 16:293–299.

Terrace, H.S. 1983. Apes who "talk": Language or projection of language by their teachers? In *Language in Primates: Perspectives and Implications*, J. deLuce and H.T. Wilder eds. Springer-Verlag, New York, N.Y. pp.19–42

Terrace, H.S. 1985. In the beginning was the "name". *American Psychologist* 40:1011–1028.

Thorrington, R.W. Jr. and C.P. Groves. 1970. An annotated classification of Cercopithecoidea. In *Old World Monkeys: Evolution, Systematics and Behavior*, J.R. Napier and P.H. Napier eds. Academic Press, New York, N.Y. pp. 629–647.

Tilson, R.L. 1981. Family formation strategies of Kloss' gibbons. *Folia Primatologica* 35:259–287.

Trivers, R.L. 1972. The evolution of reciprocal altruism. *Quarterly Review of Biology* 46:35–57.

Trivers, R.L. 1972. Parental investment and sexual selection. In *Sexual Selection and the Descent of Man, 1871–1971*, B. Campbell ed. pp.136–179. Aldine Press, Chicago, Ill., pp. 136–179.

Trivers, R.L. 1974. Parent-infant conflict. *American Zoologist* 14:249–264.

Tutin, E.G., M. Fernandez, M.E. Rogers, and E.A.Williamson. 1992. A preliminary analysis of the social structure of lowland gorillas in Lupe Reserve, Gabon. In *Topics in Primatology Vol.2.*, N. Itoigawa, Y Sugiyama, G.P. Sackett and R.K.R. Thompson eds. University of Tokyo Press, Tokyo, Japan. pp. 255–266.

Vandenbergh, J.G. and W. Post. 1976. Endocrine coordination in rhesus monkeys: female responses to the male. *Physiology and Behavior* 40:216–227.

Vedder, A. 1987. Report from the Gorilla Advisory on the status of *Gorilla gorilla*. *IUCN Newsletter* 8:58–62

Walters J R. 1987. Transition to adulthood. In *Primate Societies*, B.B. Smuts, D.L. Cheney, R.M. Seyfarth, T.T. Struhsaker and R.W. Wrangham eds. University of Chicago Press, Chicago, Ill., pp.358–369.

Walter, J.R. and R.M. Seyfarth 1987. Conflict and cooperation in primate societies. In *Primate Societies*, B.B. Smuts, D.L. Cheney, R.M. Seyfarth, T.T. Struhsaker and R.W. Wrangham eds. University of Chicago Press, Chicago, Ill., pp. 306 317.

Wasser, P.M. 1987. Interactions among primate species. In *Primate Societies*, B.B. Smuts, D.L. Cheney, R.M. Seyfarth, T.T. Struhsaker and R.W. Wrangham eds. University of Chicago Press, Chicago, Ill., pp. 210–226.

Wasserman, E.A. 1984. Animal intelligence: Understanding the minds of animals through their behavioral ambassadors. In *Animal Cognition*, H.L. Roitblat, T.G. Bever, and H.S. Terrace eds. Laurence Erlbaum Associates Publishers, Hillsdale, New Jersey. pp. 45–60.

Whiten, A and R. Byrne. 1988. The Machiavellian intelligence hypothesis: editorial. In *Machiavellian Intelligence. Social Expertise and the Evolution of Intellect in Monkeys, Apes and Humans*, R. Byrne and A. Whiten eds. Clarendon Press, Oxford, U.K. pp. 1–10.

Whitten, P.L. 1987. Infants and adult males . In *Primate Societies*, B.B. Smuts, D.L. Cheney, R.M. Seyfarth, T.T. Struhsaker and R.W. Wrangham eds. University of Chicago Press, Chicago, Ill., pp.343–357.

Wilson, E.O. 1875. *Sociobiology: The New Synthesis*. Harvard University Press, Cambridge, Mass.

Wolfheim, J.H. 1983. *Primates of the Wild: Distribution, Abundance and Conservation*. University of Washington Press, Seattle. Wash.

Worlein, J.M., G.G. Eaton, D.F. Johnson and B.B. Glick. 1988. Mating season effects in the mother-infant conflict in Japanese macaques (*Macaca fuscata*). *Animal Behaviour* 36:1472–1481.

Wright, P.C. 1992. Primate ecology, rainforest conservation, and economic development: Building a national park in Madagascar. *Evolutionary Anthropology* 1:25–33.

Wynne-Edwards V.C. 1962. *Animal Dispersal in Relation to Social Behaviour*. Oliver and Boyd, Edinburgh.

Yamada, M.K. 1957. A case for acculturation in the subhuman society of Japanese monkeys. *Primates* 1:30–46.

Zeller, A. 1987. Communication by sight and smell. In *Primate Societies*, B.B. Smuts, D.L. Cheney, R.M. Seyfarth, T.T. Struhsaker and R.W. Wrangham eds. University of Chicago Press, Chicago, Ill., pp. 433–439.

n = footnote